Prison on the Prairie

Paul Gale

Order this book online at www.trafford.com
or email orders@trafford.com

Most Trafford titles are also available at major online book retailers.

Print information available on the last page.

ISBN: 978-1-4251-0713-0 (sc)

Trafford rev. 01/07/2022

www.trafford.com
North America & international
toll-free: 844-688-6899 (USA & Canada)
fax: 812 355 4082

Dedication

This book is dedicated to my wife, Jo Anne, whose perseverance and hard work made it possible.

Foreword

By I. Ronald Lawfer

During the course of one's life, certain issues arise to become memorable events which consume part of your life and touch the lives of numerous others. They start fairly routine, but then progress into significant status. These events can be frustrating, entertaining, rewarding, challenging and educational. To me, the events covered in this book were all of these things.

What happened during the years from 1994 to 1998 are covered in this book. The author of this book, Paul Gale, covers the history of the Savanna Army Depot. Most of the book covers the closing of the depot by the Army, efforts by the local redevelopment authority to provide employment, the siting of a maximum security prison at the depot and the events which caused the prison to be moved to Thomson, a location about 20 miles south of the depot site.

For ten years, I had the honor and privilege to serve as state representative in the Illinois General Assembly in Springfield, representing about 100,000 people in five northwest Illinois counties, including Jo Daviess and Carroll counties. When my term ended in January 2002, I chose not to seek reelection. While removing files from my office, I discovered a large amount of material regarding the prison that my assistants, Pat Esker and Mary Kay Baumann, had stored.

The information was all public record and, after a review,

I felt it should be preserved as a history of the area. The files, which I moved to my home, contained in excess of 500 pages of newspaper clippings, letters, meeting notes and minutes. It took numerous hours to place all of the data in chronological order. In reviewing these files, I was amazed at how easy it was to forget certain occurrences and became newly aware of others. All of these events were important to northwestern Illinois and especially to those affected by the depot closing.

While organizing these records, I came upon a book written by Benjamin B. Stout entitled "The Northern Spotted Owl." It was a compilation of the events of how an owl, which was becoming an endangered species, led to the slowdown of the Oregon timber harvest. Stout had a background in forestry and served on the advisory committee for Oregon state representative Liz VanLeeuwen. The bulk of the timber affected was in her district and when term limitations in Oregon forced retirement, VanLeeuwen let Stout use her files.

Stout did a good job recording the issues and the history from 1975 to 2002. The real nuts and bolts of the issue were how the environmentalists and their organizations sought to achieve their objectives using the spotted owl as a lever and surrogate to stop the timber cutting. This was brought to light by Andy Stahl, a resource analyst at the Sierra Legal Defense Fund in Seattle, Wash. At the University of Oregon Law School conference in 1988, Stahl stated, "If the spotted owl had not been involved, we would have had to invent it to save the forests." The issue was stopping the timber cutting and not saving the spotted owl.

This statement reminded me of the Army Depot situation; how environmentalists and organizations connected to the Illinois Department of Natural Resources used a little-known plant called James' Clammy weed (Polanisia jamesii) as ammunition. They incorporated the same strategy as the environmentalists: using a weed to stop a prison.

The IDNR staff and the agency's satellite commission members were used as pawns to create an almost jobless economy in the Savanna area which exists today. Besides giving inaccurate testimony in Springfield, they withheld results of studies from the public. In addition, outside environmental legal teams designed to intimidate, threaten and delay legal actions were brought in.

These environmental organizations, many of which no longer exist, and the individuals who no longer live in the area, failed to carry through on their pledges and promises. They said ecotourism from the "pristine area" would far outdistance any economical advantages of a prison complex. Today, the Savanna Army Depot, consisting of 13,000 acres, provides little or no benefit to northwest Illinois or the citizens of the state.

Acknowledgments

Tom Kocal of Lanark, publisher of the Carroll County Prairie Advocate News, served as a valuable resource. His encouragement to preserve this history was very valuable. He attended many meetings and helped review the final draft of the manuscript. He also wrote a statement outlining his views of the story, which appears at the end of the book. The original files and other notes used in this book will be stored at the Savanna Prairie Advocate offices for those interested in further research.

Sharon Cholewinski of rural Elizabeth, who served as administrative assistant to the Jo Carroll Local Redevelopment Authority between 1996 and 1999, provided notes, observations and participated in the final review of the manuscript. Her encouragement and attendance at many meetings were valuable to the efforts to produce this book.

"As the LRA office took shape, excitement and enthusiasm became part of our goal to succeed in bringing economic opportunities to the Savanna Army Depot. Many people worked unselfishly bringing ideas to the table, and many other people worked 'selfishly' bringing chaos and confusion to the LRA plans and meetings. At the time, as the administrative assistant, it felt like the LRA would be negotiating in good faith, steadily moving along and then 'bam,' we would be blindsided. We were the Pearl Harbor of the Savanna Army Depot, and the environmentalists were the Japanese who bombed us," Cholewinski said.

This book and its historical perspective would not have

happened without the efforts of author Paul Gale and Jo Anne Gale, his wife and editor. Jo Anne's Epilog appears at the end of this book. When I contacted Paul, a former newspaper-man about the book, he was enthusiastic because he grew up in Savanna and his parents, a grandmother and an aunt had worked at the depot. His dedication and professional man-ner will make for enjoyable reading to those who were part of this history and to those who want to learn about these events. Paul has been successful in his interpretation of the files and the interviews in a fair and balanced way so that the reader can learn and in some cases relive the historical events.

Introduction

Beneath the Hanover Bluffs in Carroll and Jo Daviess counties in northwestern Illinois lies some hotly contested bottomland. After the Savanna Army Depot was ordered closed in 1995, developers and former Gov. James Edgar wanted to build a state prison there. Environmentalists objected, saying the land was a rare sand bar prairie which hosted more than 40 endangered and threatened species, including bald eagles, cormorants, great egrets, river otters, bobcats and the James' Clammyweed.

The weed is now held in contempt by people in the depot community who wanted to create jobs to offset jobs lost with the depot closing in the economically depressed area.

People in several towns, including Galena, noted for its tourism, also objected to the prison because they felt it would adversely affect their businesses. Both the Galena and Hanover councils voted to oppose the prison. Savanna residents and most of Carroll County desperately wanted the prison because the jobs it would create would replace the 421 jobs lost with the depot closing.

If there was to be a prison, opponents argued the state should put it in Carroll County, instead of a 140-acre site near the Whitton Gate in Jo Daviess County. Edgar believed the threat of lawsuits by the environmentalists would hold up the project, so he decided to build the 1,800-bed maximum security prison north of Thomson. Located about 20 miles south of the original site, the $140 million Thomson prison was built in 2001, but stood empty for five years because the state ran out of funds to open it. The prison was the state's second most expensive building, next to the Thompson Center in Chicago. The state legislature agreed to fund a partial opening during its 2006 spring session. Guards were trained during the summer and the first inmates were brought in after Labor Day of 2006.

Several years before the depot closed, a local redevelopment authority was formed to find new jobs at the depot. Almost immediately LRA members fought off the U.S. Fish and Wildlife Service, which wanted more than 11,000 acres of the 13,000-acre depot for a game refuge. During negotiations, the Wildlife Service settled for 9,854 acres for Lost Mound refuge. Most of it was bottom land and forest. The U.S. Corps of Engineers received 456 acres - 280 acres for recreation and 176 acres for dredge disposal.

Much of the former depot land is fenced off from the public because of contamination created by unexploded ordnance and the U.S. Department of Agriculture, which dumped pesticides at the depot years ago. So far the Army has spent $200 million on the cleanup, but the process has been slow and another $70 million is needed to make the depot safe for public access.

The contamination and Army red tape also have slowed development of the LRA land. Only 77 employees work for businesses now at the depot, as many of the proposed developments failed for one reason or another. Several went bankrupt. Besides a warehouse operation and a rail car maintenance facility, new plans call for a data server farm and a winery.

The contamination made much of the former depot dangerous for public use. The public won't for many years be able to hike or cross-country ski across the prairie, taking in the subtle beauty of the prairie along with fantastic views of the river. The Fish and Wildlife Service envisioned wide use of the land for hunting, trapping, fishing, boating, hiking and photography. The National Audubon Society had once shown an interest in establishing a migratory bird education center on USFWS property.

Many believe the nearby Mississippi Palisades State Park is so beautiful and grand, it should be joined with the Army depot to form a national park. It could serve millions of people in the nearby Chicago area; but those in the know, believe the U.S. Fish and Wildlife Service would never surrender its land and open it up to the public on the same level as a national park.

Since the six years the depot closed, about the only thing built for the public is a lookout above a majestic stretch of sand dunes on the Mississippi River banks. Some say the environmentalists who fought so hard to preserve the wildlife and prairie have now abandoned it and are seldom heard from. Several key leaders in the prison opposition have sold their homes in the Hanover bluffs and moved away.

At the time of the prison fight, the Friends of the Depot wanted to build a visitor's and interpretative center complex which would include an information desk, classrooms, educational exhibits, meeting rooms, a gift and book store, food and vending, restrooms, hunting, fishing and camping supplies, canoe and bike rental and educational and interpretative trails. None of these projects have been built, although Alan G. Anderson, USFWS refuge operation specialist, said plans are to open up the River Road another three miles to the former Coast Guard boat ramp. It is Anderson's highest priority to have the ramp opened.

Volunteers from the Natural Area Guardians, Audubon Society, Prairie Enthusiasts and Friends of the Depot have helped the USFWS clean up some of the land. More than 500 tons of scrap, 14 miles of railroad track, four buildings, three railroad loading platforms and a railroad loading dock have been removed. Eventually, all the buildings will be removed in the USFWS area but the earthen and concrete bunkers will probably remain.

Several research projects funded by USFWS are ongoing at the depot. Research is being done on the fragile prickly pear cactus, while another is studying the seed bank in soil and how the plants have responded since the grazing stopped. Part of this study involves the James' Clammyweed. There are 47 endangered and threatened species at the depot and during one count, more than 700 American bald eagles were counted last year. There are also eight eagle nests at the depot.

Terrance Ingram, president of the Eagle Nature Foundation and co-chairman of the depot restoration advisory board, doubted whether the boat ramp can be opened anytime soon because in June, shells were buried in the ground about a foot deep along the road.

The clean up is not proceeding "fast enough' to suit Ingram, an Apple River resident, as the federal government is only funding about $10 million a year for the work while $30 million a year is needed. "There's got to be a lot more clean up done before it is opened (to the public)," he said.

The depot was declared a Superfund site due to the presence of the old burning ground. A volunteer who helps several times a year, says that it still hasn't been cleaned up. Buoys warning boaters to stay away were placed in the Crooked Slough partly caused by the burning ground. "The EPA is very careful, not so much because of health issues, but because they don't want any lawsuits," Ingram stated.

About 30 trained specialists are working at the depot with the clean up, and agents of the state and federal environmental protection agencies show up at the depot once or twice a month. Ingram felt both EPAs are being run too much by politicians.

Depot started in 1917

Chapter 1

Formerly farm land, 13,062 acres of the "sand prairie" were purchased by the U.S. government on July 23, 1917, for an U.S. Army installation.

The main gate of the former army depot is about eight miles north of Savanna, bordered on the western side by the Mississippi River, with the Apple River to the south and Blanding Landing recreation area to the north.

The 14-mile-long property, ranging from one to four miles wide, was purchased by the Army for $67 an acre as a proof and test facility for cannons manufactured by the Rock

Island Arsenal. One of Carroll County Review Publisher Jon Whitney's ancestors sold his farm, a quarter of a section which was mostly sand, to the government at a good price. He invested the money in a Hanover bank only to lose it when the bank president absconded with all the bank's money. The bank failed shortly after that and other depositors who sold their farms to the government lost their money, too, Whitney explained.

The depot construction began in April 1918 and officially opened Dec. 26, 1918, as the Savanna Proving Ground. Proof firing of weapons had begun earlier, in September of 1918. (1)

The Army started expanding the site almost as soon as it opened. Three barracks, each with a dining hall for 100 men, and quarters for commissioned and noncommissioned officers were built. Forty-seven standard magazines, 30 high-explosive magazines, a combination field office, storehouse, additional railroad facilities to handle movement of ammunition, and a nitrate storage pit were added.

When World War I came to a close, the gun testing diminished and the site was converted and expanded to store artillery vehicles, trucks and tanks from the war. Forty artillery storage warehouses were built and utilities and roads were expanded. Fifteen buildings remain from the original construction.

The 52nd Ordnance Ammunition Company was assigned to the Savanna Proving Ground late in 1920 and remained there until the outbreak of World War II. Operations expanded to include ordnance storage facilities and loading and renovating shells and bombs. The installation's mission changed to a depot facility in 1921. The proving grounds became the Savanna Ordnance Depot with the depot becoming independent of the Rock Island Arsenal.

Transportation for workers from Savanna to the depot was provided by a special work train operated by a contractor. (2) Temporary bunk houses also were built for all the laborers who wanted to live there. Board was provided by the Western Boarding and Supply Co., operating under a subcontract. In a schedule of wages approved in 1918, brick masons made 81 cents an hour, carpenters 67.5 cents an hour, painters 55 cents an hour, watchmen, 32.5 cents an hour, skilled laborers 40 cents an hour, plumbers 75 cents an hour and well drillers $7 a day.

A shell-loading operation was built in 1931 to load and renovate 155-millimeter shells and 300-pound bombs. During World War II, additional facilities were constructed, and the depot was given the task of studying and developing methods to stabilize ammunition for shipping. The War Department authorized construction of 407 igloos; 26 smokeless powder magazines; 55 standard ammunition magazines; a clipping, linking and belting plant; a shell loading plant; 14 warehouses; and a generating plant. All were completed during 1942 and 1943 with the Group One shell loading plant the only one of its kind in the U.S. This plant was used to load bombs for Gen. James Doolittle's historic raid on Tokyo in 1942. (3) Eight thousand civilian employees at the depot helped send the special greetings to Hirohito.

A Chicago Daily News reporter, during a 1943 tour, said women played an important role at the depot. "The influence of women is more pronounced here than at other installations which newspapermen were privileged to inspect this week in a tour conducted by the 6th Service Command to show how the Army supply services are functioning in wartime. Our escort through the shell-loading plant was Edith Trusloe, 32, of Bellevue, Iowa. Formerly working at a large New York City bank, Trusloe was now supervising the activities of hundreds of employees who handle TNT as though it was a melted candle, which it resembles," the Daily News said. (4)

Like most shell-loading plants, Savanna assembled component parts made elsewhere, packing the complete rounds in fiber cases and packing these into large wooden crates, most of which were trundled out of the buildings on to waiting railroad cars to be shipped to combat units. Live ammunition waiting shipment was given code names.

Matches and spark-producing metals were forbidden in the shell and bomb loading areas, and production was stopped during electrical storms when employees retreated to underground shelters. Floor sweepings were carefully burned as safety measures. Newspapermen were peering down into a 500-pound bomb body, watching TNT being packed, when a fire alarm sounded. Employees scurried to designated shelters while the reporters searched frantically for an exit, wishing the $10,000 government insurance they purchased three days earlier at Camp Grant was in effect.

A similar alarm was sounded the previous year when champion heavyweight boxer Joe Louis visited the bomb line. Louis was tipped off ahead of time, but one of his assistants ran down the line scared he would be blown up. According to the Daily News, three minutes after the alarm, as part of a practice drill, three fire trucks arrived at the gate, sirens wailing. Hopping off the trucks, the first firefighters seemed like any other skilled hose handlers, but a closer glimpse revealed feminine eyes peering from beneath the rubber helmets.

"The time isn't far off when 70 percent of our employees will be women," commanding officer Col. M.A. Brackett predicted. "We've lost 1,350 men to the draft so far and will soon lose a similar number. The only trouble with women is that they shy away from supervisory jobs. They seem to dislike the task of bossing other women."

An appeal went out in October 1944 for more women to work because the depot wanted to double its production. Brackett said more bombs and shells were needed by the Army due to the "increased demand for supplies made by our combat troops in their smashing attacks and drives into enemy territory."

"There must be many women in the towns and on the farms in this territory who are able to leave their homes and concentrate their efforts on aiding in winning the war. How nice it will be for these women to be able to say 'I did my part' when their soldier returns after victory," Brackett said. Recruiters were ordered to go house to house in Carroll County to contact women to work.

One of the female workers was Bertha Miller, my grandmother, whose husband Matt, my grandfather, farmed in Bureau County. While Grandma worked at the depot making small ammunition, she lived with her daughter, Dorothy Gale, my mother. As a secretary in the Ammunition School, Gale could live in Black Hawk Village, a residence for depot employees. Grandma left the depot as soon as she had saved enough money to buy an electric stove. My mother worked there for several years.

Besides the women, Italian prisoners incarcerated at the depot were enlisted to aid the war effort. One of their jobs was reclaiming shell cases by knocking out primers from British shell cases. "These men now are anti-fascist, and are said to be eager for an early Allied victory so they can return to their homes in liberated Italy," the Clinton Herald reported on Aug. 5, 1944. The Italians also loaded and unloaded trucks and railroad cars and crated ammunition. Besides being paid the regular rate of pay, the Italian workers were given additional freedom.

"This takes the form of sight-seeing or educational tours and recreational visits to the nearby communities under military supervision. This relaxation of the restrictions has resulted in a noticeable improvement in their efficiency and productiveness in the war effort. Members of the Italian Service thus have a different status than ordinary prisoners of war, such as German prisoners and Italian-Fascist prisoners. This arises from the fact that Italy has been accepted by the U.S., Great Britain and Russia as a co-belligerent in the war against Germany," the Carroll County Mirror-Democrat reported. Savanna residents remembered seeing buses taking some of the Italian workers to the movie theatres in town.

The Italian prisoners allowed to work were investigated to make sure they were not pro-Nazi or pro-fascist. "If they pass this screening test they are classified according to skill or aptitude, given such training as may be necessary and assigned to duty in essential war work ... Although many members of the units have requested combat duty, it is not allowed under the terms of the Geneva Convention to use them in combat," the paper said.

Brackett agreed to let a Dubuque reporter and photographer come to the depot in August of 1944 in an attempt to clear up rumors in Dubuque that the Italians weren't getting along with civilian workers. "There has been no friction and the work by the Italians has been satisfactory," Brackett told newsmen. (5)

On the tour, the newsmen saw Italians loading freight cars with war materials (not ammunition) destined for overseas. Another group was salvaging a large quantity of lumber used to ship ordnance to the depot. The lumber was used to build crates to ship ammunition. A third group was packing copper oil containers left over from WWI, while others were reclaiming brass in a mountain of shell cases sent back from overseas. Italians also could be seen building a wire fence, a project delayed for months because of a lack of manpower. Two of the plants where the Italians worked were busy around the clock.

The Italians awoke at 5:50 a.m. and worked from 7 a.m. to 4:30 p.m. Those working close enough to the Italian mess hall marched there for lunch. Those working too far from the mess hall were served on the site. They had their own barracks and own cooks.

The newsmen sat in on a lunch at the Army officers' mess where several Italian officers were feted as guests. "The gathering was conducted in the manner of a service club meeting with peppy songs ... Col. Brackett made the announcements then the proverbial piano player appeared and the group including the Italians sang such tunes as 'Hail, Hail, the Gang's All Here,' 'Old McDonald Had a Farm' and several others equally disastrous to the artistic ear - not mine," the Dubuque reporter wrote.

The highlight of the singing session was the appearance of an Italian officer tenor and an American officer with a baritone voice which was reported to have been heard at the Metropolitan Opera in New York. They sang two numbers and won ovations on both occasions. The Italian promised to sing a solo at the next Rotary meeting. "The incident is mentioned to indicate proof of the contention that the Italians and Americans are working and playing together in a spirit of genuine friendliness," the reporter wrote.

One of the Italian officers, who spoke five languages including Arabic, was captured by the British in North Africa. He had been a prisoner in England and Canada before being transferred to Savanna. Another Italian smiled broadly when newsmen suggested the day was hot and humid. Spending 22 months in the camel corps in Africa, he said the temperature soared to 150 degrees one time. The Italians remained in holes dug in the ground during the hot days and came out only at night.

Another officer said he could identify the planes which flew over their African territory by the number in the group. For instance, if one plane flew over, it was German; if there were two, they were British; and when the sky was filled with planes, they were Americans.

Most of the Italian officers spoke English and some fluently, but few of the enlisted men spoke English. This deficiency was solved through the use of interpreters. The Italians wore American uniforms described as "castoffs" and "seconds."

Earlier in 1941, 37 buildings of temporary construction were erected at the depot as part of a unit training center. With quarters for 945 enlisted men and 40 officers, the complex had 15 temporary barracks, five mess halls, six recreation buildings, officers' quarters, a post exchange, a guard house, two storehouses and an administration building. Most of these buildings are still standing.

Along with the building expansion, the number of depot employees mushroomed from 143 in 1939 to 7,195 in 1942. To help meet the housing shortage, the Federal Works Administration in 1941 built a 200-family housing complex called Craig Manor in Hanover, seven miles north of the depot entrance.

During 1943, the Federal Public Housing Authority constructed another housing project called Blackhawk Village about ½ mile east of the main gate. Blackhawk, where I lived until age four, was turned over to the depot in 1948 and sold to the city of Savanna in 1976. Most of the buildings are now demolished. In addition, 12 dormitories (six now remain), three recreation buildings (two now remain) and a mess hall, which has been demolished, were constructed within the main gate.

Operations included munitions making, renovations, testing and disposal, firefighter training, land filling, use and disposal of solvents, fuels and pesticides.

During a speech to the Clinton Lions Club in September of 1944, Army Staff Sgt. John Vasos of Clinton told the group he flew 23 missions over Germany in a Flying Fortress. Vasos said he was pleased to drop bombs made at the depot on the Germans. The Jan. 17, 1945, Clinton Herald ran a photo showing depot employee Alyce Smith of Savanna signing her name to a 100-pound bomb she hoped would be dropped by her son, Lt. Robert Swingley, a bombardier navigator, on the Germans. Employees at a bomb-loading plant who exceeded their quota during a bond drive got to sign their names on bombs.

The depot phased down after the war, but hundreds of underground bunkers remained. On Jan. 21, 1948, an igloo-type magazine containing 150 tons of TNT exploded, breaking windows in Hanover and Savanna and was felt over three states. Nearby magazines were seriously damaged, but no one was injured. The igloo contained anti-tank mines. The force of the concussion opened locked doors, shook furnace pipes loose and shattered several large plate windows on Savanna's Main Street. The blast was felt as far south as Springfield and northeast in Lake Geneva, Wis.

Savanna residents reported their floors jumped, after which there was an instant of silence and then a terrific roar. Buildings and houses in Morrison shook, and there followed rumblings like that of thunder. While there were no reported injuries at the blast site, two Hanover Woolen Mills employees were injured slightly by flying glass from broken windows at the plant. The windows dated back to before the Civil War.

Hanover village board member John Eberhart was out for a walk and said the blast almost knocked him off his feet. Across the river in Green Island, Iowa, the Rev. Lewis Ryan was plastering walls at his church. "I first checked the oil stove that heated the building to see if it had exploded, and then ran outside thinking it might have been a nearby farm home. We saw a glow in the sky from the direction of the depot." Later during an open house at the depot, Ryan passed near the 100-foot by 50-foot-deep crater. "It really was a big hole," he exclaimed.

Long distance operators were swamped with calls from inquisitive people. Many thought there had been an earthquake. A Chicago railroad official called Rockford authorities asking whether there had been an earthquake because he feared rails were damaged. The explosion occurred only a few minutes before a Burlington Zephyr passenger train passed over the tracks which run through the depot grounds. The tracks lie about two miles from where the igloo exploded. A railroad station a half mile away from the blast was undamaged, but telephone poles and lines were knocked down across the tracks. Rail traffic through the depot was disrupted for nearly two and one half hours before the damage was repaired.

Special troops from Fifth Army headquarters in Chicago were dispatched to the depot to relieve depot guards and members of the state police.

Florence Walker, who lived in a farm house near the depot, said the explosion blew the doors and windows from her house. She was so frightened she ran into her yard to see what was happening, and saw the sky was filled with fireworks, followed by a huge cloud of smoke.

The Rev. Alphonse Schmidt, head seismologist at Loyola

University in Chicago, reported a slight disturbance on the university seismograph.

Within minutes, streets were jammed with traffic as motorists wandered aimlessly around looking for the source of the explosion. Bedlam reigned in police stations as scores jammed the offices and attempts were made to contact nearby communities by shortwave radio. Gates to the depot were closed shortly after the blast and admission was denied to all. Under this veil of secrecy, rumors spread thick and fast. In Clinton and the adjoining area, the rumors ranged from an attack by an enemy nation to the accidental explosion of an atomic bomb.

The igloo contained 50,000 anti-tank mines, sent to the depot for storage after WWII. Testimony in Army files showed it took nearly three-quarters of an hour before workers located the blast site. Parts of the cement structure were found as far as three-fourths a mile away.

"They never found the doors," said depot employee George Delp, years later.

High-flying concrete chunks from the igloo caused extensive damage to other structures, forcing an expensive removal and rebuilding program which lasted nearly six months. Although no estimate was given for the damage, depot employee John A. Gerdes later told a Dubuque Telegraph-Herald reporter it was in the tens of thousand of dollars.

A four-month investigation found no official cause. "It was felt unofficially that the blast was caused by some malfunction of the mine itself and not the storage procedure," Gerdes said. Depot safety officer Roland Unangst said that a faulty fuse on an anti-tank mine caused the blast. Afterward, depot employees removed fuses from all the land mines in nearby igloos and pulverized them in a rock crusher.

Two factors apparently aided in containing the damage.

One was the design of the concrete-encased igloo. The igloos were designed to allow any blast to throw its force in a guided direction away from other igloos. Also, the amount of dirt and foliage on top of the igloos kept the force of the blast from spraying the concrete chunks farther.

Depot employees excavated dirt from a site to maintain a two-foot earthen cover over the igloos. The site where the earth was taken was later called Primm's Pond. It is named after one of the employees who dug up the earth. Primm's Pond is one of the "pristine" sites the environmentalists later wanted to preserve.

Another igloo exploded in July of 1949. It contained chemical fuses, and the blast was limited to inside the structure's walls. Depot officials called it a "fizzle" instead of a blast. "There was no evidence of shock, due to the small amount of explosives in the magazine," explained Harold G. Evans, public relations officer at the depot.

Having grown up in the upper valley of the Plum River, former state Rep. Ron Lawfer, a Republican dairy farmer from Stockton, remembers it was not uncommon in the late 1940s for the valley to get cloudy on a hot summer day, the result of the depot testing smoke bombs.

A significant event for the depot occurred in 1950 with activation of the Ordnance Ammunition Surveillance and Maintenance School. It was renamed the Army Material Command Ammunition School in 1966 and provided technical, operational and administrative training in all ammunition fields for civilians and military students from the U.S. and foreign countries.

The Korean conflict resulted in increased hiring and building at the depot. The War Department decided to make the depot one of the largest ammunition storage bases in the country. During 1954, residents from nearby communities complained about explosions when the Army destroyed obsolete ammunition. An estimated three million pounds of combustibles and explosives were exploded and burned in 50-pound packages. "We're doing the job as economically and safely as possible. We've been at this job ever since the end of World War II, but not until this fiscal year (from July 1, 1954 to July 1, 1955) have we had enough money to hire the people to really clean up the situation. We're making a real hole in our stock of dangerous munitions," reported depot commander Col. Leland A. Burbank. (6)

Complaints and fears were voiced by residents, particularly in Bellevue, Hanover and Savanna, where atmospheric conditions and shock waves seemed to be close. Burbank had words of assurance for them. "Utah University mining engineer students have made studies that prove that 1,000 pounds of explosives detonated on hard ground will do no damage to a normal house, a mile away. We're exploding only 50 pounds at a time on sand and we're 3.1 miles from the base of the Bellevue dam and Bellevue across the river is the closest town."

Burbank acknowledged that when heavy, cold air and low clouds bounced sounds and shock waves down toward the surrounding towns, residents' nerves are upset. "I know folks think we've got quite a conflagration, especially when clouds hang low, but it's nothing at all dangerous - just pretty."

When asked by a reporter why the explosives weren't shipped down the Mississippi and dumped in the Gulf of Mexico, Burbank said it wouldn't be safe. "Every time you handle this old stuff, you're putting another nail in your coffin. We can burn or explode it in small, safe quantities. We couldn't afford to ship it in small quantities. Working with explosives is like driving. Some drivers don't have a safety instinct; some workers don't have it. We don't employ people without that instinct - They wouldn't have time to correct their mistakes on this job."

A special weapons mission, which included nuclear weapons, was started at the depot in 1961 and continued through 1975.

During its heyday, the depot had a budget of $48.4 million a year with between $1 million and $3 million awarded to local businesses for construction contracts. It was estimated the depot workforce comprised 48 percent of Carroll County and 21 percent of Jo Daviess County with many other workers commuting from Iowa.

The depot, renamed the Savanna Army Depot in 1962, was busy during the Vietnam War as many bombs were shipped by rail. The site was all but finished as a depot following the war when depot facilities were moved toward the coasts.

According to Jo Daviess County Board member Don Crawford of Hanover, who would later be a Local Redevelopment Board member, the depot was responsible for the proper way ammunition and missiles were blocked and braced when placed on rail cars, trucks or ships for shipping.

The depot's largest tenant was the Defense Ammunition Center established as the Army's Material Command Ammunition Center in July of 1971. It was renamed the U.S. Army Defense Ammunition Center in July of 1979.

Rumors about secret commando training at the depot cropped up over the years. An event in 1971 only added to the rumor fodder. Counter Intelligence Corps sent a crack five-man team of invaders to test the depot security and were able to penetrate the depot at two points, according to a depot newsletter called Dep-O-Gram. (7)

"An alert (depot employee) Blackie Bowman led to the

capture of three of the agents at the Washout Plant in short order. They had gained entrance by simply walking across the dam. The other two agents, far more inventive and who weren't detected, planted a simulated explosive device in the basement of Building No. 1."

"One, posing as an Army officer, pretended a need for medical attention and was routinely cleared through the front gate and escorted to the medical facility. While there the fifth agent, who had been secreted in the trunk of the pseudo officer's automobile, let himself out and made his way over to Building No.1 where he was joined by the fake officer. Unchallenged as they calmly toured the building, they selected the site for planting the simulated explosives, then turned themselves in, revealing their true identity," the Dep-O-Gram said. There were many times when the agents could have been challenged.

"On a key military installation such as this, security has to be everybody's job, and anyone who finds humor in this series of events is completely out of touch with reality. Blackie's actions were highly commendable, but then it was no less than what you would expect from a man who has always displayed a total commitment to his job and to the depot. As for the others, we can all be most thankful it was merely a test exercise," the newsletter printed.

Nuclear weapons were stored at the depot under heavy armed-guard by a special MP unit. Vehicles were searched both entering and leaving the special weapons area. Mirrors were used to look underneath the vehicles. One time, an electrician said guards had their .45 caliber pistols in their hands as they watched him repair an electric lock to an igloo door. Another time, a depot employee was wrestled to the ground by guards when he appeared to be doing something suspicious.

The nuclear weapons were removed in 1975, flown by Huey helicopters to the Chicago-Rockford International Airport to a base in California. The mayor of Rockford heard of the nuclear weapons about a month later and raised a ruckus, said Carroll County Review Publisher Jon Whitney. (8) He was upset with the government flying over his city without notifying him. "That was the government's way. They don't tell anybody anything."

"They weren't fused. They couldn't have exploded because the triggering device wasn't armed. That happened the same time we lost the nuclear weapons division up here. I was more upset with that than the other ... those were high-quality jobs," Whitney added.

"As the base phased out of activities, especially special weapons, it was somewhat on a regular basis that my wife and I were awakened about 4 a.m. by black helicopters, generally in groups of three, to the northeast. I and everyone else seemed to know one was loaded and the other two were decoys," Lawfer said. (9)

Depot Col. L.A. Burbank told the Savanna Chamber of Commerce in 1954 there was little chance the depot would be the target of a nuclear attack. "Few targets are worth the great cost of an atomic blast," Burbank said. Bridges and railroads in the area would be more important targets, though not, of course, atomic missile targets, he said.

During a 1959 Armed Forces Day, visitors to the depot were shown a liquid-propelled Corporal missile, called the nation's first true ballistic missile. Being replaced by the more-advanced Sergeant missile, the Corporal was capable of striking targets beyond the range of conventional artillery. The Savanna chamber learned in 1959 that rocket fuel also was stored at the depot. "There's no cause for alarm, for these mixtures are harmless babies while separated ... The guided missile field is getting bigger at the depot and we know we will be getting more of this work," assured Col. Albert C. Wells Jr., depot commanding officer.

Wells remarked atomic warheads probably will be stored at the depot, "but if we were storing them, you would never know about it."

Simulated atomic explosions were the highlight of the 1965 Armed Forces Day. Three detonations of atomic simulators were seen by many of the 3,000 people attending the event. Each detonation brought a sharp explosion, fire ball and a huge mushroom cloud. Visitors were told the atomic simulator was a safe substitute for the atomic bomb and was utilized in training troops.

In 1959, the depot employed 825 people and had a payroll of $385,000 a month. During the Vietnam War, more employees were hired. By the close of 1965, 143 military and 911 civilians were employed, and by 1967, the civilian force climbed to 1,270. In November 1967, the depot was directed to reduce its civilian force to 1,042 by the end of the year. Hardest hit by the reduction was the depot's maintenance and administrative areas. Gov. Otto Kerner came to the depot to help it celebrate its 50-year anniversary on July 22, 1967.

Even though nuclear weapons were removed in 1975, students at the ammunition school continued to receive training in maintaining nuclear weapons and missiles. Some of the school's courses dealt with handling, storage and transportation of nuclear weapons, said a 1974 booklet prepared by the U.S. Army Material Command Ammunition Center. (10)

Alpha monitoring team members were trained to control radioactive contamination and to prepare firefighting teams to deal with a radiation or an explosive hazard situation. The

1974 booklet contained pictures showing students training on Lance and Redstone missile propulsion systems.

Although the Department of Defense recommended the depot close in 1976, it got a reprieve eight months later. At that time, there were 413 civilians and 15 military personnel at the depot. Savanna Mayor Don Nehrkorn and Hanover Mayor Don Hunt co-chaired a "Save Our Depot" campaign. Nehrkorn said their efforts paid off. Depot commander Lt. Col. Richard Bogenrife stated an Army study concluded the cost of closing the depot was "so great that it could not be recovered through the recurring savings within a reasonable time."

U.S. Rep. John Anderson, a Republican from Rockford, estimated the cost of the closing at about $250 million. Nehrkorn praised the work of Sens. Charles Percy, Adlai Stevenson of Illinois, Dick Clark and John Culver of Iowa and U.S. Reps. Anderson, Tom Railsback of Illinois and Mike Blouin of Iowa for convincing the Army to keep the depot open.

In November of 1974, the Army cut the depot work force from about 1,000 to 428 when the special weapons division was transferred to the Sierra Army Depot in California. Remaining at the depot were the 259th Ordnance Detachment, the U.S. Army health clinic, the ammunition center and school and the Army Communication Command detachment.

With a mission to recycle no longer-wanted ammunition, the Demil (demilitarization) Technology Office was established at the depot in February of 1993. Ordnance and large rocket motors were transported to the depot where they were recycled or burned.

Also on the premises, sits an old stone house used by the Underground Railroad during the Civil War. The house and several depot buildings, including the C area igloos, were nominated for inclusion on the National Register of Historic Places. (11) The National Park Service said two WWII buildings were of interest because they were prototype buildings for numerous other WWII buildings. The igloos were described as the first igloos of their design, and the last survivors of their design.

The old stone house was built in the 1830s and is one of the oldest structures in Jo Daviess County. For many years, it was used as a guard station when the depot was patrolled by horseback. Those on patrol would ride the 15 miles from the stable near the main entrance and would then stay overnight, returning the next day.

According to the "Carroll County, a Goodly Heritage," the old stone house was built by John Beaty who used it as a stagecoach and pony express stop between Galena and points south. Stone for the house was quarried from what is now the basement. Before the house was built between 1836 and 1840, the acreage where it stands was a tract of wild land over which the feet of white men had scarely passed.

Named Beaty Hollow after John Beaty, the homestead consisted of 323 acres and included the site of the present store at Blanding. During the early days, a ferry was used to cross the Mississippi River at the present site of the Bellevue dam.

"It was a common custom to hold corn shucking bees in the fall of the year moving from homestead to homestead until all the corn of each farm was harvested. Upon completion of the corn picking homesteaders often would hold a party for neighbors who helped. In the case of the stone house, Beaty held dances upstairs. In 1845, John Beaty's father and his two brothers joined a sheriff's posse. They were sent to arrest an outlaw named Brown and his gang of horse thieves and counterfeiters at Bellevue, Iowa, just across the river from their house. Four of the outlaws and four members of the posse were killed in the fight. The remaining members of the gang were captured and the next day, according to legend, were tied to a tree and each given 49 lashes.

Brac 95 Commission rules against depot

Chapter 2

During 1995, the depot, with 420 military and civilian employees, was designated to close under the Base Realignment Closure (BRAC) Act. At the time of its closing, it had 923 buildings with an annual payroll of $16.2 million. On Feb. 28, Secretary of Defense William Perry said the Savanna Army Depot Activity was cited by BRAC 95 to close. Savanna Alderman Jack Fosdick hated to hear the announcement at a time when he felt Savanna was "starting to make a comeback."

The closing will have a "tremendous impact on Jo Daviess and Carroll counties," lamented Hanover Mayor Don Schaible, who worked at the depot for ten years. Hanover was especially hit by the closing when earlier its Eaton Corp. plant had packed up and moved to Mexico.

Larry Straight of Hanover, who worked at the depot for 27 years as a civil engineer and technician, believed he could get a job with a contractor after the depot closed. But the closing would make contract jobs more scarce because local contractors had done so much work at the depot. For instance, during 1994 a Dubuque construction company completed a munitions-testing facility which cost $5 million. Other employees might have an even tougher time finding employment. "There's not much of a job market for the type of work accomplished down here. Most is real specific to the depot," Straight remarked.

This wasn't the first time the depot's future was threatened. In November of 1974, the Department of Defense announced plans for a large phase-down of depot activity, and in 1976 the Department of Defense recommended the depot close its doors. Community leaders and the U.S. senators and representatives from Illinois and Iowa convinced the Army to change its mind.

Once the base closes, the Army will help military and civilian employees find new employment or develop new skills, promised Major James E. Sisk, depot commander. "Any closing would be done with sensitivity to the people working on the base and to the community ... Our infrastructure must be reduced proportionally with our military forces. The people who work on this installation have provided excellent support to the Army and this nation for over 77 years and will continue the tradition as long as they are required to do so." (1)

As in 1976, politicians and members of the community rallied against the move to close the depot. On Feb. 16, 1995, Gov. Jim Edgar appointed Lt. Gov. Bob Kustra to head a multi-agency task force to defend Illinois military bases likely to face scrutiny by a federal commission.

"Military facilities located throughout Illinois have played a critical role in the nation's defense. It is important that we have a cost-effective national defense ready for the next century, and that will require base closings throughout the nation. But before any bases close, we want to make sure that a fair and accurate assessment of Illinois' contributions is made and that military spending is equitably spread among the states. We will do whatever is necessary to save Illinois jobs from unfair or misguided downsizing of the military," assured Edgar. (2)

Federal, state and local officials met March 3 at the depot to begin developing a strategy to deal with the closing announcement. "The state of Illinois plans to put the best case forward to save the depot and at the same time acquire help from other state agencies to find a use for the depot," Kustra announced.

Kustra, who had not visited the depot prior to the meeting, said he "had no idea how much was up here." Looking at a topographical rendition of the depot he remarked that Illinois and the area might be much further ahead if the depot was closed. He observed that the site offered considerable economic benefits for new and expanding corporate ventures, and added, "maybe we should be asking, 'how soon can we get them out of here?'"

The closing of the depot would result in 421 lost jobs and have a dramatic impact on local retail and service businesses and on housing costs. "Whenever you take that many high-paying jobs out of a rural area, the economic impact will be felt," said William Ernst, acting director of the U.S. Army Defense Ammunition Center and School, which was scheduled to move to McAlester Army Ammunition Plant in McAlester, Okla.

All of those working at the ammunition school have the option of transferring to McAlester. "The employees on the depot side weren't as fortunate ... If the school relocates, the area will lose a lot of good people, professionals who contribute to their communities," Ernst added.

Several depot employees were former servicemen. Bob Hanson, a depot employee for 15 years, groaned, "I got the job at SADA because I was a Vietnam veteran. When it closes, I'll be a Vietnam veteran without a job." (3)

"It's something we have been thinking about, but the reality is devastating. We lost the railroad in 1980 and are now just starting to recover," said Savanna Mayor Eugene Flack.

Also affected by the projected closing would be the area school districts. In March of 1995, the Savanna School District had 68 students, who had parents or guardians working at the depot. The district received $9,909 in Federal Impact Aid for each student. "We will do an analysis to see how many people will relocate and how that will affect our enrollment figures," Savanna School District Superintendent Paul Seymour told the Times-Journal.

U.S. Rep. Lane Evans, a Democrat from Rock Island who represents the area, said he would fight to keep the depot in Savanna. "I plan to testify as to the importance of the depot and I'm disappointed with the announcement," Evans remarked from his Washington, D.C., office.

Both employees and surrounding communities will be eligible to receive assistance under BRAC Sisk said. Economic assistance will be available to private or public businesses interested in locating at the depot after the closing,. Sisk declared, "We're going to do everything we can to make the possible transition as easy as possible, not only for our employees, but for the entire community."

"This town's taken several other hits in the past and the world didn't come to an end," said Steve Haring, who was president of the Savanna Chamber of Commerce and an advertising salesman with the Savanna Times-Journal at the time. He would later serve as executive director with Jo-Carroll Depot Local Redevelopment Authority, a group formed to redevelop the depot. "We seem to be a resilient community and people. And in the long run, whether we are fortunate to have it stay or it is gone by 2001, we will rise from the dust."

"You just don't believe it will happen. When I heard it was a very strong possibility, I was just sick," lamented Ginnie Jansen, owner of Wirth's, a jewelry and antique shop. Even with the closing, Jansen said the town could survive. (4.) "Savanna has a lot of tourism. Naturally, it will hurt, but my business doesn't depend on the depot."

Townspeople told the Sterling Daily-Gazette they had seen the writing on the wall for many years since the depot was downsized from 1,000-plus employees. Depot civil engineer Straight questioned why the Army was shutting the depot when it spent millions of dollars there during the past 25 years.

John Morehead, administrator of the Fort Sheridan Commission from 1989 to 1992 and future LRA executive director and board member, said, that in the long run, the local economy could benefit from the closing. "The areas surrounding other bases that have closed have bounced back. It takes a little time, but over a period many have again as much or double the employment as before. There is a history of a silver lining."

The depot developed a transition team designed to help employees. "We're not going to wait. We are working together as a team to develop a plan to handle the BRAC decision," Sisk emphasized. "We're going to do everything we can to make the possible transition as easy as possible, not only for our employees, but for the entire community."

Lawfer, a Republican dairy farmer from Stockton, wrote U.S. Sen. Carol Moseley-Braun on March 30 asking for support to save the depot. "I would point to three areas that I believe have not been part of the decision-making process to close the Savanna Army Depot. This facility has in its favor one of the most expansive rail systems available on site which provides access to a nation-wide rail system. It also has a barge capacity with respect to the Mississippi River. Combine with that, the unique and extremely well-maintained U.S. Army Defense Ammunition Center and School buildings, which were recently built to operate that school, I would submit to you there is increasing evidence that the Savanna Army Depot has greater value than originally estimated."

Moseley-Braun wrote back saying the BRAC Commission uses three categories of criteria to evaluate the Secretary of Defense's decision to close bases. First, the commission assesses the military value of each base, then measures the return on investment of closing bases. Finally, it evaluates both the economic and environmental impact of the base closure on the community. "In the cases of the Savanna Army Depot and the Charles Melvin Price Support Center (near Granite City in southwestern Illinois), I do not believe that the facts support the closure of these bases. I do not believe that the criterion of the BRAC process have been met." (5)

Lawfer also received several letters about the closing from his constituents.

"The economic impact this closure will have on Carroll County and Jo Daviess counties in Illinois, and Jackson and Clinton counties in Iowa, would be devastating. In addition, I think if the formation this decision was based on was more closely looked at, I think you would find misinformation and mistaken assumptions which seriously affected the decision-making process," wrote Carl E. Mutters of Hanover.

R. R. Larson, area manager of Interstate Power Co. in Savanna, wrote Lawfer asking him to put the fight to keep the depot open at the top of his priority list. "The Save the Depot Committee feels that there may be other more cost-efficient alternatives for the government to look at rather than closing of the Savanna Army Depot."

A task force was formed to present a case of economic duress and economic loss to the area. They met for the first time on March 19, 1995. When BRAC commissioner S. Lee Kling visited the depot on April 11, 1995, 75 people at a hearing were told he would "look, listen and learn" as part of an independent process to investigate the issues critical to the base.

For another hearing on April 12, 1995, two busloads of area residents traveled to the Rosemont Convention Center in Chicago, to voice opposition to the closing.

Elizabeth Mayor Lynne Hesselbacher complained the bus ride gave her motion sickness, but she had to go because 25 Elizabeth residents' jobs were at stake.

Awake and ready for the bus, Carl "Skip" Schwerdtfeger of Old Northwest Realty was used to getting up at 4:30 every morning. "I'm a taxpayer and I want our government to be efficient. It's not efficient to close the depot," he told a Dubuque Telegraph-Herald reporter. He explained it would be easy for the Save the Savanna Army Depot Committee to talk about the impact on the local economy, but it chose a better route - showing the commission that the Army could save money by keeping it open.

On one of the buses, Hanover real estate agent Bill Wolter remarked "sometimes I think those people want to hear from good, down-to-earth citizens who will speak from the heart, rather than from someone who will put on a show."

The depot contingent had only 30 minutes for its presentation, and that time included opening remarks by Kustra and Moseley-Braun. The task force told the BRAC panel the Army greatly underestimated the military value of the depot and the economic impact the closing would have on the communities.

Moseley-Braun testified that environmental costs to clean up the depot would be astronomical.

"Although the Department of Defense says that it is obligated for the costs to clean up all bases and does not factor environmental costs into the decision to close a base; in reality, Savanna may never be able to house a commercial tenant," Moseley-Braun asserted.

Karen Stott of Savanna told the BRAC Commission the closing would increase unemployment by 2.8 percent in Jo Daviess and Carroll counties.

Back home, Savanna was decorated in a "Save the Depot" motif. Red ribbons covered homes and businesses and signs saying "Save the Savanna Army Depot and USADACS" sprouted in yards throughout the small city. "We are not accepting it. It would be devastating if we lost it. It really would. Losing those people - because they're the key people in the community," cried Jane Smith a Savanna resident.

The late Albert Ehringer, director of the U.S. Army Defense Ammunition Center and School and former civilian executive at the depot, advised the commission closing the depot would create a national ammunition storage problem.

Keeping the depot open would save the government $100 million, he added.

A discrepancy in information used by the Army to rank the ammunition depots it was considering for closure was probably not large enough to change its recommendation that the Savanna depot close, a General Accounting Officer evaluator explained in April. (6) The GAO released a report on the process used by the Department of Defense to select 146 bases for closure or reassignment. The Pentagon's work was "generally sound and well-documented," but it revealed in an evaluation of eight ammunition depots, it found problems that "in some cases, raise questions about the reasonableness of specific recommendations."

Despite the task force's hard work, the BRAC commission agreed with the Department of Defense's decision to close the depot. The official announcement came on June 23, 1995. "After serving the U.S. Army for the past 77 years, only three minutes were needed Friday to close the Savanna Army Depot Activity," the Savanna Times-Journal reported. The eight-member commission voted 7 to 0 to eliminate the depot. U.S. Sen. Alan Dixon, the commission's chairman, abstained from voting, citing a conflict of interest.

Savanna residents got "a raw deal," U.S. Rep. Evans complained. (7)

"We've got to go through the agony of watching this closing for the next five, six or seven years, without knowing if it's even possible to develop it," predicted Haring.

The atmosphere around the depot was tense that day. "I would say everybody's a little on edge," concluded Ron Johnston, a depot electrician. When the depot shuts down, Johnston, 52, said he will look for another civil service position. If that fails, he'll try the private sector. "But who's going to hire someone in their 50s when they could get someone younger?" he asked.

Haring believed the pain that the Savanna area was feeling echoes throughout the country, as other bases close. "But we'll go to bed tonight, and the sun will come up tomorrow. It's a hard pill to swallow," he remarked.

Barney Barnhart, acting chief of operations, felt closing the depot makes no sense. "We have a very high efficiency rate. The number of tons we shipped during Desert Storm was comparable with that of bigger depots. We've never missed a shipment. We have a great reputation." He called the decision personally devastating. "I'm 45 years old with a daughter starting college this fall. This is the last thing I expected when I came here in 1983," he grumbled. Barnhart calculated that the average age of the depot workers was 47.8 years old, too young to qualify for full retirement benefits by the time the installation closes but perhaps too old to find as good a job.

51-year-old Rich Hartman, a painter-worker who was employed at the depot 30 years, observed that by the time his work at Savanna would be phased out, he might be 52 or 53 and would need to be at least 55 to get full retirement benefits. He had no idea what he would do.

"It will hit the town hard. We've already lost every industry we have here," cried Virginia Elliott, a waitress at the Town House Restaurant in Savanna. (8)

The Chicago Milwaukee and Pacific Railroad yard closed in Savanna in 1985. It had been one of the largest in the Midwest. Now it resembles a newly planted forest on the South side of town sprouting small trees and weeds.

"Savanna doesn't have a lot going for it right now. God only knows what will happen. We will have a lot of homes for sale," sighed Carol Dunnigan of Savanna, who worked at Red's Antique House. "It's kind of sad. My dad, he helped start the depot in 1917 and he worked there for years. I lived in one of those white houses there until I was six, when my mother died."

Dave Davies, visiting pastor for the First United Methodist Church in Savanna, predicted the church stood to lose 50 people. "That's about 15 percent of our congregation. It will make quite a bit of impact on Savanna. It will have quite a bit impact on the whole region."

The closing recommendation was forwarded to President Clinton, who passed the list on to Congress. The U.S. House later voted overwhelmingly to accept the closure of 79 military bases. The depot's closing would be started within two years and finished within six years.

When Helen Schamberger of Stockton heard news about the depot closing, she called the office of Republican U.S. Rep. Don Manzullo of Egan to see whether they were aware of it. "They said just a minute. When they came back on the line, they told me to call Lane Evans' (a Democrat from Rock Island) office because the base was in his district. This really irritated me to say the least. It didn't take me long to set them straight," she declared. Manzullo's office learned three quarters of the depot was in Manzullo's district. "How sad. I did call Lane Evans' office and they were well aware of the closing and even knew how much was in his district," Schamberger added.

Later in an interview, Whitney expressed his belief if John Byrd Jr., who headed the ammunition school, were still alive, he would have kept the depot open. Byrd died during the winter of 1995 when he was on a tractor removing snow from the driveway of his rural Savanna home. Whitney felt Byrd died of suspicious causes, but sheriff's reports do not substantiate his theory.

"John Byrd was an extremely talented individual. His

civil service level was comparable to a three-star general," Whitney said. "John told me one time, (when another fight succeeded to keep the depot open) 'I'm not really doing this for Savanna. If we weren't competitive and it couldn't do the job, and it was cheaper to do it elsewhere because of costs ... I wouldn't keep it here. But it is significantly cheaper to keep it here. The costs of doing business here is so much lower, the cost of housing for everybody." Byrd believed the Army shouldn't put everything in one basket, a tendency to put everything on the coasts. "He thought they should spread things out."

Meanwhile, Lawfer learned in February 1995 the depot was included in the federal Base Closure and Realignment Commission for closure. Also during February, the Army announced it was moving the Defense Ammunition Center and School to McAlester Army Ammunition Plant in McAlester, Okla.

"In March several groups and organizations were formed to save SAD (Savanna Army Depot). Some former employees of the depot provided ... financial reports so that some of the cost allocated to the SAD really belonged to other institutions. I along with a bus of others from Savanna and Hanover traveled to Chicago to hear input to BRAC 95 commissioners. The result, of course, was that the base would close," Lawfer explained.

The closing would have a special impact on the Bill Robinson family as members of three generations worked at the depot. (9) Robinson was chief of the mission division at the depot. His grandfather and both parents worked at the depot during its 77-year history. He was planning to pass the tradition down to two of his children who spent summers as lifeguards at the depot pool. "My mother and father met at the depot during the war. I started here in 1966 right out of high school. It's been home to all of us." Robinson wasn't surprised about the decision to close the depot because of its small size. He also hadn't decided what he would do after 29 years at the depot, although he might try for a job at the Rock Island Arsenal, or take early retirement.

Years later, Haring declared that not enough was done by the Illinois U.S. Senate and congressional delegation to keep the depot open. "I just had a flat out no. That's the only way I can state it."

Ehringer told the Quad-City Times he was upset to learn that Defense Secretary William Perry was revising the base-closing list to benefit a California military installation. (10) Ehringer asked if President Clinton could make an exception for the nation's most populous state with 54 electoral votes. "Why couldn't he make an exception for Illinois with 22 electoral votes? We are in an unholy position of being out here in the sticks. I've never seen anything more political than this."

The Perry compromise was intended to defuse a politically charged debate over the closing of the McClellan Air Force Base and the loss of 11,000 military and civilians jobs. California stood to be hit hard by the base closings. McClellan, near Sacramento, was the largest of six military bases in California the BRAC Commission recommended closing. The Perry plan was designed to keep half of the jobs in the area by allowing the Air Force to hire private companies to do the base's maintenance work.

Ehringer, pointed out the same shortsighted closure plans occurred after World War II when ammunition storage facilities were closed. "This was a very, very large mistake. We all realize we've got to cut back. But you don't cut back on your war reserve assets. I thought we learned our lesson from the past, but we haven't." Ehringer called closing the depot "a national disaster" and asked for President Clinton to appoint an independent group to "investigate the whole ammunition stockpile" issue. (11)

The government's need for ammunition storage and demilitarization was overwhelming and the need for the depot was as great as ever. In terms of national security and national interest, the depot closure would compound the problems of ammunition storage worldwide. "It's a catastrophe that it's being closed," predicted Ehringer, who participated in a local study which found serious flaws in the decision to close the depot. Asked about future uses for the depot, Ehringer believed it was doubtful that any industry would be interested in the area because of the contamination problems and liability associated with it.

LRA rises like a phoenix out of the ashes

Chapter 3

Now that the depot was scheduled to close, a local redevelopment authority organization began to form.

During a meeting Aug. 15, 1995, ten members were named to a newly formed Savanna Army Depot Local Redevelopment Authority executive board. Chosen were Gene Mauldin, Mayor Eugene Flack and John Sullivan, all of Savanna, Jim Rachuy of Stockton, Joel McFadden of Shannon, Jack Viviani of Thomson, Don Crawford of Hanover and

Hanover Mayor Don Schaible, Penny Lauritzen of Lanark and Preston Duncan of Camanche, Iowa, a member of the Mesquaki Tribe. The board members were selected from a nominating committee among a pool of 34 applications and were declared elected by consensus.

Following the action and a round of applause, Col. Ron Adkins declared it was the first time an LRA board was chosen this way. It was usually done by a ballot. After Sullivan, owner of a grocery chain, was named the board chairman, the executive board appointed a grants writing committee to begin work on writing a grant to secure initial funding from the government.

During a previous organizational meeting, Don Wanatee, member of the Mesquaki Tribe from Tama, Iowa, requested the Sac and Fox Indians of Iowa be included in the LRA. His tribe was interested in one-third of the depot and intended to create a "spiritual center" for the tribe. His group also planned to develop a resort, museum and interpretive center for the five tribes from the Great Lakes area.

Wanatee discussed development of a recreation area which did not exclude a gambling casino. The Mesquaki Tribe operated a casino in Tama and other tribes throughout the country are involved in similar casinos. "People are pretty much natural gamblers. I'm not an expert on economics, but I think on-shore gambling is coming to Illinois. I think everyone in Illinois probably knows that."

Wanatee admitted he does not believe gambling offers a solid basis for economic development. "Everything comes and goes in cycles. Gambling will be the same way. I wouldn't say you could base economic development on gambling because gambling too often takes from the very people you're trying to help." Wanatee believed a Mesquaki center on the depot would benefit the surrounding area, but was uncertain if northwestern Illinois residents would share that belief. He thought American Indians had a primary claim to the area, because their ancestors lived there for centuries before the land was taken over by whites.

Meg Bushnell, representing the Illinois Department of Natural Resources, noted part of the depot was a national forest until 1952 and said federal agencies had one of the first rights of refusal of any Department of Defense property slated for closure.

The newly named Jo-Carroll Depot Local Redevelopment Authority held its first meeting on May 21, 1996. During a discussion of the board's legal status, Carroll County state's attorney Val Gunnarsson, newly appointed board member, declared, "We are an agency of two counties, a unit of government. We are a government entity."

The ten members were equally divided between Carroll and Jo Daviess counties. At the Jo Daviess County Board meeting on May 13, County Board Chairman Judy Gratton appointed Schaible, Crawford, Rachuy, Bob Wehrle of Galena and John Rutherford of Apple River to the board. All those appointed by Gratton, except for Rutherford, had served as members of the LRA planning agency. Rutherford was involved with the Jo Daviess Natural Area Guardians and served as a liaison between the county board, of which he was an elected member, and the county zoning committee.

Members of the new LRA board from Carroll County were appointed by Carroll County Board Chairman Bill Ritenour. They were Gunnarsson, Sullivan, Carroll County Clerk Judy Gray of Chadwick, McFadden and Flack. Gray and Gunnarsson had not served on the LRA planning committee.

The board agreed Haring, who was hired by the planning board as LRA executive director in January, would stay on as permanent executive director. The board also elected Sullivan as chairman and hired Sharon Cholewinski as LRA administrative assistant. Opening an office above the Blackhawk Area Credit Union in Savanna, the LRA set a goal of attracting a variety of economic activity to the site.

"We have 3,200 acres for the LRA for the reuse plans. We hope to have light industry and possibly clean, heavy industry, vocational-technical training and administration and small community businesses," Haring elaborated. (1)

Haring also raised the possibility of a boot-camp style prison. "The DOC (Illinois Department of Corrections) is still interested in possibly siting a prison complex." With three and one half miles of shoreline at its disposal, Haring told the Journal-Standard that high-end housing would nicely fit into the mix. Having 1.25 million square feet of space, the project had unlimited possibilities. "We will have an open forum in January (1997) to discuss the prison issues. We want feedback, and if the climate is right, we will continue to pursue (the prison)."

Haring told the paper a consulting firm would hold public hearings later in 1996,when a reuse plan for the depot becomes more detailed. He encouraged people outside Savanna to get involved.

"This is a regional effort, it's not just about Savanna."

Col. Ronald Adkins of the U.S. Office of Economic Adjustment announced to a group of 50 local civic and business leaders on July 14, 1995, that the LRA must speak as one voice for the community. (2) "The Army, economic adjustment office and all other agencies involved in the closing will deal with only that one entity, the LRA. Period." Adkins explained the economic adjustment office would provide technical and financial assistance to the group.

Speaking as one voice would not be easy, Haring told the

group. "My biggest concern is meshing the area as a whole. The three counties have never really had to work together. How do you get everyone thinking in the same frame of mind and finding common ground?"

Not all the land would be used for a single purpose, the Dubuque Telegraph-Herald reported, as it will likely be parceled off for many uses, such as parks and recreation, local industry and education.

LRA member Rutherford noted he would like to see the northern half of the depot, located in Jo Daviess County and much of which was covered with Mississippi River islands and sloughs, used for recreational purposes. With access to fishing, duck hunting and wildlife, more tourists would be attracted to the area. More tourists would bring more jobs. "In the long run, it will create more than 400 jobs (the number employed at the depot), maybe not in a dramatic way, but with things like bait shops, boat-rental places and overnight accommodations."

Rutherford didn't see any point in trying to lure industry to the area. He predicted Carroll County would establish an industrial park of its own in the southern half, which would make the competition for industry even greater.

Wanatee felt his community had an interest in preserving ancestral burial sites, located on the base and in the surrounding hills. He proposed a museum and interpretative center be built. The site had potential for fish and mussel harvesting, plus it could serve as a tourist destination with a resort and amusement park. "Long before this was an island or an Army installation, it was home to the Mesquaki Tribe. The Sauk and Fox Treaty of 1804 gave the land to the tribes. Our villages were here. There are burial sites up on the bluff. We lived here a long time before this was an Army depot," Wanatee said. (3)

Preston Duncan of the Mesquaki Tribe reported his tribe originally wanted the entire property. "We had thought we'd put in a recreation area, with different sections for different recreations. We had wanted to create a lot of jobs, more than are here now. Some areas would have been religious areas. There are places on this land that are sacred to us."

In a March 3, 2006, letter to Sharon Cholewinski, Duncan explained the depot property was once the home and center of the Mesquaki Nation, numbering more than 2,500 residents in the latter part of the 1600s. Smaller bands lived on both sides of the Mississippi River, from north of Prairie DuChien in Wisconsin to St. Louis. "Knowing my history, I have a very strong spiritual ties with those areas and the people. In 1993, as I walked on the hills and bluffs surrounding the depot, I began hearing rumors about the depot's closing. In 1995, when it began to happen, I began to realize we had a chance to better the economy for both the Mesquaki (Nation) and the pilgrims of Savanna," Duncan wrote.

Earlier, Duncan began talking to the Tribal Council hoping to get them interested in the depot property. "A very small group was interested, but they had no governmental power. I was made a speaker for the tribe by the Sac and Fox Tribal Council (in 1993)," he wrote. When Duncan was picked to be an LRA member, he remembered there were people who voiced negative opinions about Indian involvement. "These elements didn't concern me since they were from Dubuque. As time passed by, I began to learn my situation and I realized we had those negative elements right in key positions that were supposedly acting as coordinating elements between the Army and the LRA. It was this element that kept the Mesquaki interest away. He told the tribe that if they took the land, they would have to clean the land up before they could move on it," Duncan wrote.

No matter what the obstacles, Duncan kept trying. "My first interest was to bring an international free trade zone to Savanna. It was ideal because of the railroad yard in Savanna, the airport and the highway system. Also the waterway could be utilized. Because of the helipad, I wanted to build a hospital there on the base with an attached nursing home for paraplegics and learning center for the deaf with rehabilitation in mind. I had tourism in mind, of giant canoes traveling up and down the Mississippi with Indian villages on the islands and along the banks, with very big pow wows, maybe three times a year. Learning centers of the Native Americans, their cultures and languages in mind. I also wanted to bring a hideaway and resting spot for government officials both foreign and local. Everything was feasible. Even (if it required buying more land). I was even interested in buying the bluffs to utilize them for hiking trails," he reaveled.

The LRA experience was a great honor for Duncan. "It was very much like a beautiful dream I refused to release a hold of. I was a volunteer during my involvement. I was never employed by the Sac and Fox tribe of the Mississippi in Iowa. This status gave me a freedom to work with other tribes if they were interested. I notified the Oklahoma Sac and Fox, the Kansas Sac and Fox, the Kansas Kickappo, the Wisconsin Ho Chunk and Chippewas. I only received a brief interest from the Ho Chunk and a brief response from the Minnesota Sioux," Duncan wrote.

During the summer of 1998, Duncan made one last effort to regain some depot land, but it proved to be futile. He was invited to work with a company, Electronic Recovery Specialists, Inc., and moved to the depot. The venture didn't work out, and Duncan lost about $3,000 in exercise equipment when another company claimed it while he was gone.

Wanatee suggested the property be divided three ways.

Fish and Wildlife would have a third, the U.S. Forest Reserve a third and the Mesquaki Tribe the rest.

The corps had plans to use its area as an outdoor recreation site with campsites. "There will be additional campsites, a trail and access to the Bellevue Dam. They have Apple River Island and a place they will use a river bottom dredge site," Haring announced. Negotiations at the federal level took longer than expected. "We need to continually push the process. We kind of got bogged down in the federal (phase). But, it was not a negative experience. We got an education and learned a lot," Haring explained.

Formation of the LRA board was a BRAC requirement, and there was heated discussion over how to divide the property, Crawford revealed. (4) "In the end a compromise was achieved, and the property was divided as it is at the present time. The next requirement was to hire an engineering firm to complete a reuse plan and submit it to the BRAC office. The plan was completed and submitted, however, as with any reuse plan it has been tweaked and revamped several times."

In August of 1995, the Mesquaki tribal council gave the go-ahead allowing the 1,300-member tribe to take part in planning the depot's redevelopment. The tribe operated a gambling casino in Tama, Iowa, and many in Savanna believed the tribe wanted to open another casino at the depot. If the Mesquaki tribe wanted to establish a casino without going through the screening process, it would have to get the gaming activity accepted as part of the base reuse plan, which must be endorsed by the community and the governor, said Michael Chapman of the Bureau of Indian Affairs.(5)

The Mesquaki Tribe would have a difficult time establishing a claim on the depot, observed David Etheridge, a lawyer in U.S. Department of Interior's solicitor's office which advises the Bureau of Indian Affairs. Asserting such a claim through litigation would be next to impossible because the statute of limitations expired a long time ago in the Mesquaki case he explained. On rare occasions, Congress has passed legislation establishing ownership if a tribe can show that land it rightfully owned was improperly taken. "But that's a fairly lengthy process and not a very promising one."

Even though the Mesquaki Tribe missed a deadline on a filing date to acquire depot land, Haring proclaimed "we were willing to work with them. They came and went. For whatever reason, things didn't happen there." Recently, Haring discovered that another representative from the Mesquaki Tribe in Oklahoma again investigated opportunities at the depot. But, when talking to the LRA board executive director Dave Yllinen, he learned the tribe didn't follow up on their inquiries. "They kind of came and kind of went again. They just disappeared. It's one of those things. It's frustrating because we tried to work with Native Americans on some things. It didn't work out," Haring added. (6)

In January of 1997, a reuse plan and implementation strategy was prepared for the LRA. At that time, the plan called for 9,445 acres, including 6,000 acres of bottomland, to be transferred to the U.S. Fish and Wildlife Service, 3,157 acres to the LRA and 460 acres to the Corps of Engineers.

Quest for a state prison

Chapter 4

One of the first moves by the LRA board was to ask Illinois Gov. Jim Edgar to chose the depot as a site for a state prison. The state was in the process of building 15 prisons during the last 16 years, and it was rumored that if another prison was built in Southern Illinois, the state would tip on end due to the weight of the prisons. The deciding factor offered by the Illinois Department of Corrections was the area's unemployment rate and desire by the community to have a prison in its area.

Edgar was urged to select a Northern Illinois site, ruling

out the eastern part because no community near Chicago ever applied for a state prison.

A critical factor in the Illinois Department of Correction's decision was community support for a prison. An IDOC official commented "we're not going to put a prison where it is not wanted, but on the other side of the coin we have never placed a prison where there has been 100 percent support."

Edgar had the final word on the site and local unemployment figures would be an important factor in the decision. An IDOC news release on Sept. 14, 1996, noted 32 communities had submitted applications for the 1,800-inmate prison. A Galena resident also submitted a bid, but after a visit to the site, IDOC felt it did not meet the needs of the department.

On Sept. 23, IDOC officials toured the Savanna Army Depot visiting potential sites for a prison. Arlen Dahlman, base transition coordinator, told corrections officials about the infrastructure in place and said the areas proposed were above the 100-year flood plain. Dahlman added all of the sites would be checked for environmental significance.

"I firmly believe the Army is willing to work with us for redevelopment of the area on a fast track if we are to get something like this in," LRA board chairman John Sullivan emphasized. Competition for the new prison was heavy and Beardstown offered to construct the prison for 15 percent less than DOC's estimate. "It's hard to compete against numbers like this. But, we are and I think we've got a good chance."

On Oct. 2, Sullivan testified at the Illinois Valley Community College's Cultural Center in Oglesby for Savanna to be a prison site. "We have four different 100-acre sites at the Savanna Army Depot, all of which have the potential to be ideal sites for your new prison. We invite you to select any one of these locations and we will give it to you free." (1)

The depot would be pleased to share electrical, water and

sewer service available at the depot with the Department of Corrections. "We can also assure you that we can have the natural gas for you as soon as necessary ... Our area offers an abundance of high quality, well-educated individuals in the work force, which you will need. With the foreclosure of the depot, 421 jobs will be lost and all of those individuals have well-developed skills and trades. We have security guards, electricians, plumbers, machinists and accomplished office and clerical workers," Sullivan testified.

When Sullivan moved to Savanna in 1967, it had a population of 5,200. In 1995, the population dropped to 3,800. "You can see that we have had a significant decline in population, which has been a direct result of cutbacks with the Savanna Army Depot, the railroads, closing of small industries and the youth camp (which closed Oct. 31, 1975)," he added.

Sullivan explained that besides the loss of 421 jobs, a Northern Illinois University study found 224 spin-off jobs would be gone with closing of the depot. The area could lose $16.8 million a year in personal income, and lost economic activity might total $35.5 million.

On Oct. 10, Savanna was picked as one of five finalists for a $65 million medium-security prison. The other four finalists were Marshall, White County, Decatur and Pinckneyville. (2) "These look like places we want to put prisons," declared IDOC spokesman Nic Howell.

In a letter to Edgar, state Sen. Todd Sieben, a Republican from Geneseo, wrote that Savanna was the most attractive package for a prison. (3) "Public support for this site has been overwhelmingly positive. The most frequently heard comment from the public is 'this location really makes good sense.' It will also help attract other business and industry to the site," commented Sieben, in the letter co-signed by eight other area lawmakers.

At an Oct. 12 hearing at Savanna, most of the those attending voiced overwhelming support for the prison. (4) Besides the points made at the Oglesby hearing, Sullivan testified in Savanna that the depot also offers a firing range on the base for security guards to use and has a "beautiful home that overlooks the Apple River that would make an excellent warden's home."

There are eight adult correctional facilities in southern Illinois, 12 in central Illinois and six in northern Illinois. "We believe that to be geographically balanced this site fits into the plan for the state," Lawfer told the 200 in attendance at the hearing.

Gene Mauldin of the LRA reported a recent informal survey found more than 98 percent of the people surveyed support a prison in Savanna. Among the two percent was Patricia Kennedy, the only person to speak against the prison. Kennedy asked why the prison would receive the land free and why others couldn't have the same deal.

Kevin Swan hated seeing family members having to move away to find jobs. "I'm tired of missing family members at Christmas and Thanksgiving," he moaned.

John Thompson, president of the Dixon Chamber of Commerce, answered concerns about a prison in his town since he said he lived within "spitting distance" of the Dixon Correctional Center, a medium-security prison. "It is not stopping people from investing in residential property." Retail businesses have seen the benefits of visitors and the local payroll. In addition, there have been no security problems. The load on local officials and agencies has not increased significantly," reported Thompson. He added that prison jobs pay well and have excellent benefits.

Many at the hearing asked questions about the impact of the families of inmates. The families do not come to Dixon to ask for housing, public aid or public health services, Thompson inferred. After their discharge, IDOC Director Odie Washington said 95 percent of the inmates return to their original communities. Both Washington and Thompson said families come into a community, visit a prisoner and return to their homes.

The Rev. Greg Albert, formerly a chaplain for a prison in Virginia and now Savanna Presbyterian Church minister, assured the crowd that the Virginia prison had no impact on tourism and inmates' families did not move to the area. Members of the Virginia prison staff were "good citizens" who were active in the community as leaders and as volunteers. The prison also would offer the community an opportunity for spiritual outreach, Albert noted.

Curt Eubanks of Lanark, a ten-year employee of the Dixon prison, told those at the hearing of the economic benefits, career opportunities and chances for personal development available to prison employees.

Savanna alderman Bill Scott believed his neighbors have less to fear from a prison than the high explosives stored for decades at the depot.

Many in the crowd applauded when Tom Robbe, who worked at the depot, commented, "We've lost a hospital, two railroads (Burlington and Milwaukee), the ice house, Eaton Corp., Garment factory, Helle's Lumber Co. and now we have 13,000 acres at the Savanna Army Depot. Do we make it into another Milwaukee Railroad 'jungle' and receive no revenue from it?"

Former Savanna resident Kevin Swan, who now lives in Hanover, was undecided about a new prison before coming to the hearing. "After listening to the presentation I'm excited about the prospects of a new prison," concluded Swan.

Longtime Savanna community leader Carl Lantau

was born on the depot. "The site is in nobody's back yard. Nothing but positive things could happen from this. I'm in favor of it," he declared.

Jo Daviess County Board Chairman Judy Gratton told Washington that sewers and roads will be built to accommodate the prison. "We'll do what it takes to get you to locate here."

Although natural gas was not available to the site, an Interstate Power representative advised there's a ready supply waiting hook-up. Steve Haas of Interstate Power informed the group the utility can provide electricity to the prison at four cents per kilowatt hour, which is the lowest rate in the state.

City officials in prison towns contacted by the Quad-City Times agreed that the positives of a prison outweigh the negatives.(5)

East Moline Police Chief Gary Sutton called having a prison "a learning experience for the community, the prison and the police department." Sutton lived close to the prison and was a police officer when the prison arrived. There had been few problems. "I think prisons sometimes get a bad rap. People say the prison brought the gangs, or the prison brought the drugs. That isn't necessarily true. We do have some problems in the community, but I can't say that the prison is the cause. That's too easy, to just blame the prison. It's often a Catch 22. People want the prison for the economic base, but then they blame the prison if they have a problem with crime," Sutton emphasized.

Mayor Arlene Carlson of Fort Madison, Iowa, believes her constituents accepted the Iowa State Penitentiary, built in 1839, as a part of life. "We did have a walk-away from minimum security recently that did prove to be disastrous," she added. It that case, a family was taken hostage and a woman was raped. "The new warden and the governor immediately moved toward the idea of keeping a sex offender out of minimum security. We recently haven't had that many escape breakouts attempted," Carlson concluded.

Dixon, Ill., Police Chief Robert Short viewed the Dixon Correctional Center as a positive factor in the community, even though there has been some increase in crime. "Any time you have a prison, your crime is going to increase. It isn't anything we can't handle, and it certainly isn't something that would make me want the prison to leave," Short acknowledged. "It's been a definite asset, having the prisoners come into town to work. As a matter of fact, I have a janitor who comes in from the prison. We've had crews of prisoners painting fire hydrants or painting lines for the street department."

David Moberg wrote in an Illinois Times article "State Jobs: Economic Boon or Bust?" that nobody can blame communities for trying whatever they can to revive hard-hit local communities. (6)

"And at least some public jobs generate other lasting benefits that spread through the local economy. Public sector paychecks once were far from generous, but with the power of their unions behind them, public employees today often are as well or better paid than comparable private sector workers and have more security. Prison security guards, for example, make $26,000 a year or more and have good benefits. Because these state jobs are less dependent on the boom and bust of the private economy, cities like Galesburg seek public employers," Moberg printed. He quoted Galesburg Mayor Fred Kimble who said "they make a contribution to the economic base that helps you resist fluctuations."

Tougher sentencing laws and the war on drugs have led to a massive increase in prisoners. As a result, corrections budgets have grown much faster than other categories. "The budget grew from one percent of spending in 1965 to 5.9 percent in 1995. Yet this only accounts for money spent to run prisons, not the cost of constructing new ones. Since 1973, the state also has appropriated $1.2 trillion for building new prison capacity, nearly one-fifth of total capital spending in Illinois. The spending greatly accelerated after 1978; in the five years from 1973 to 1977 corrections claimed only 3.5 percent of capital spending," Moberg wrote.

With budgets so tight, rising prison costs compete with other uses for state funds. Due to rapid growth, the corrections department has most strikingly posed the question of the relative economic benefits of public employment in recent years. When the prison building boom accelerated during the early 80's, the Department of Corrections opened the process of prison site selection to intense bidding, taking into account appropriateness of the site for the state, local community support, and, last, economic need. There were many communities in need.

"Sometimes it gets ludicrous. One community did a video of a guy in a barrel singing 'Is you is, or is you ain't, going to give us a prison,'" Nic Howell, a spokesman for the Corrections Department, revealed.

The prison selection process is not immune to political manipulation, and critics have complained that Republican governors in Illinois have chosen sites that would be beneficial to GOP legislative interests. Moberg also noted that the state's siting process encourages communities to bid against each other with "incentive packages that are costly to already depressed towns and villages and at odds with Edgar's administration policy that, at least ostensibly, eschews use of incentives to lure new jobs to an area."

Most communities believe that the long-term benefits

will exceed the cost, though typically projections exaggerate the indirect effects. Boosters often claim that the prison payroll will ripple through the local economy, generating five or six times the number of prison jobs in other local businesses, from hardware stores to fast-food outlets. But economist Joseph Persky of the University of Illinois at Chicago argues that, on average, "the correct multiplier of economic impact for public facilities is only 1.6 to two times as many jobs in peripheral firms as in the core business," Moberg reported.

J. Fred Giertz, professor of economics at the University of Illinois at Urbana-Champaign, advised that the state "shouldn't go around squeezing areas and do like firms in pitting one area against another. It's strange, state government doesn't like that game when it's played by firms, but to a certain extent, they're doing the same."

First round goes to Pinckneyville

Chapter 5

A couple of weeks after the hearing, Gov. Edgar announced the 1,800-bed prison would be built in Pinckneyville in southern Perry County. (1) Pinckneyville offered the state a package of incentives worth $8 million. The town had been hit hard by a coal mine closing which cost more than 1,500 jobs.

"I'm very disappointed with the decision. Everybody worked hard to bring the prison to Savanna," Flack moaned.

The state passed over Savanna, Edgar declared, because it

was not clear how quickly the state could build at the depot. "Our concern was how long it would take to get that property to us," he said. Lawfer planned to ask the Department of the Army to transfer the land to state ownership, so it could be considered for future prison expansion.

State Sen. Dave Syverson, a Republican from Rockford, thought Savanna could be the site for a regional boot camp, something nine legislators in northwest Illinois had talked about. "You'll see us concentrating on this issue during the winter months," reported Syverson.

State Sen. Todd Sieben felt the experts agreed Savanna would have been an excellent location for a medium-security prison. "In addition, I believe the economics strongly supported the Savanna site. There was tremendous support in the two local communities and in the area for building the prison at the Army depot," Sieben rationalized. (2)

LRA board member McFadden said that the LRA would continue to work to redevelop the depot. "Economic development is not an event or a place, but is a never-ending journey," he stated.

Sullivan concluded that seeing so much support for the prison from area residents was a positive sign.

The Department of Corrections was still interested in depot property and several steps had been taken to "fast-track" the property into IDOC hands. Dahlman wasn't aware of any proof which could substantiate Edgar's explanation for giving the prison to Pinckneyville. Dahlman felt the DOC didn't think property transfer was a problem, but Edgar obviously did. The Army told Dahlman that a special exception would have been made for DOC to acquire the land. (3)

In December, Savanna hoped to land a regional boot camp at the prison. (4) "We have an excellent site for a minimum-security boot camp as well as a DOC correctional facility and we want them both," Sullivan remarked. A 25 to 30-acre site was on the southeast edge of the depot, which Sullivan believed could easily be converted into a boot-camp facility. A boot camp is a minimum-security facility designed to teach nonviolent juveniles discipline before they get into serious trouble. Sullivan said the depot already had the basic infrastructure in place for such a camp, including barracks, a mess hall, a gymnasium and outdoor recreation facilities.

As for another prison, Sullivan added "we understand they are going to look for another site very shortly and we want to be chosen."

At their January 1996 meeting, the LRA members mentioned they hadn't given up on the hope of landing a prison. The U.S. Fish and Wildlife Service was asking for 11,400 of the 13,000 acres at the depot and LRA members felt the request was too much. "We are alarmed at their request. It was much more than we had anticipated," explained Haring.

Besides bottom lands, backwaters and islands, Haring was aware that the USFWS wanted much of the upland area, too. As to what the LRA wished to do with its land, Haring said, "We are looking at areas for light industry and heavy industry, agricultural storage, home sites, barge and grain terminal, boat docking and launching, recreational cabins and lodging."

Jim Fisher of the USFWS told the LRA during a meeting in February that he wanted to open a dialogue with the LRA. "Please don't forget the tremendous wildlife that is there in its natural state and the tourism that could be generated by this. We are interested in serving the public, too," Fisher explained. (5)

LRA board member Rachuy, as president of Northwest Illinois Prairie Enthusiasts, wrote a letter which angered Sullivan and others on the board, because he said the LRA had proposed to the Army that the entire base be opened to economic development. (6) "Do you want these prairies and wetlands to be saved? If so, then the time for you to speak is now. Do you want to preserve this unique place for the otters, eagles and maybe even bison? If so, then I urge you to write to your representative today," Rachuy pleaded.

He went on to describe the immense natural resources on the base, including dozens of threatened species and thousands of acres of prairies, savannas and wetlands. "Neither the Army nor the LRA is much interested in these resources beyond the extent required to fulfill their legal obligations," Rachuy wrote.

In response to the letter, Dahlman replied that the Army was trying to work with the USFWS. "We are not required to do any of the studies that we have undergone. The studies include management and further use for the wildlife areas," Dahlman added. Sullivan answered saying that the LRA needs to work as a team. "Anything sent out as a representative of the LRA should be approved by the LRA board," he cautioned.

Shortly after the request, Lawfer received several letters from his constituents supporting the USFWS position. "It would be a wonderful addition to our area's heritage and for all the people to enjoy. I am particularly interested because of its historical significance," wrote Lewis Reisner of Warren.

In another letter, Barbara Rutherford of Apple River, wife of now former LRA member John Rutherford, stated Lawfer's help would be appreciated. "I totally understand the need for economic development of a portion of the 11,000 acres comprising the Savanna Army Depot. I feel this can be accomplished while preserving a good portion of the acreage for migratory bird life, endangered and threatened plants and animals and allowing study and research there."

James and Elizabeth Runchey of rural Hanover wrote "it seems that this proposal represents a once-in-a-lifetime opportunity to preserve a piece of Illinois' natural heritage. Converting this into a wildlife refuge would seem to make good economic sense ... We are all aware of the economic growth in the area that has been spurred by tourists, most of whom come here because of our unique beauty. Increasing the recreational resources of the region will surely enhance that economic growth."

Jim and Rose Goode of Galena agreed the LRA had not gotten the public involved and had not polled the public to see what it wanted done with the land. "Both the goals of preserving our natural heritage and to promote economic growth can be achieved. The Local Redevelopment Authority should have a responsibility to poll the public on this issue and to keep the public informed as to its progress ... Please do your part to serve the public and save our natural prairies, wetlands with endangered species, threatened species or species of federal concern," the Goode's wrote.

Reading several studies, Richard B. Curtiss of Stockton stated in a letter that development of the depot will alter it forever. "Do we need more of the blight we see all around us? Like the Joliet Arsenal property, we have large numbers of potential visitors very close to this property. I feel the economic benefit of preservation will be even greater than any benefit from development," Curtiss penned.

Stockton painter Lily Tolpo wrote that lots of people really care about preserving the depot. "It's a natural treasure from an economic point and a blessing from a spiritual and ecological point. New homes are going up around here because of this value."

Northwestern Illinois officials (photo above) met with Gov. Jim Edgar in March of 1996 in Springfield to discuss prison opportunities at the depot. From left were: Arlen Dahlman, depot base transition coordinator; Savanna Mayor Eugene Flack; Steve Haring, Local Redevelopment Authority executive director; John Sullivan, LRA chairman; state Rep. Ron Lawfer; and state Sen. Todd Sieben. Below, a big crowd listened to speakers during a hot evening at Stockton High School on July 17, 1997.

On Feb. 27, the Illinois Department of Natural Resources agreed with the views of many residents when it issued a press release saying the depot "is a jewel of biological diversity." (7) "With its 13,000 acres of habitat that vary from extensive blocks of Mississippi River backwater lakes and forests to vast stretches of sand prairie and savannas, this U.S. Army depot is one of the best kept secrets in the Midwest," the IDNR declared.

Two awards totaling $20,200 from the department's Wildlife Preservation Fund would be used to study the habitat, announced Brent Manning, IDNR director. Besides taking aerial photos and an extensive ground survey, the study would fund an invertebrate inventory of the prairies, keying in on butterflies and leaf hoppers. Randy Nyboer, regional administrator with IDNR's Division of Natural Heritage, agreed an accurate account of habitat types at the depot was needed with detail as to specific acreage, plant composition and natural community differentiation.

Environmental interest in the depot continued when the Nature Conservancy of Illinois wrote Lawfer on March 6 stating a modified request from the USFWS for depot land should be granted. (8) "The request of USFWS seeks transfer of only the most ecologically significant property at the depot. These lands are an ecological treasure and to permit their development, especially when ample land is available for development and near by the depot, would be a tragedy," wrote Conservancy Director Bruce W. Boyd. He noted that converting ecologically significant portions of the depot into a refuge makes good ecological and economic sense.

"Substantial economic benefits would be realized from the creation of a refuge. Over the past 20 years, much of the economic growth in the communities near the depot has been growth spurred by tourists who have come to the area to enjoy its natural beauty. People come to hunt, fish, bird watch, ride bicycles and otherwise relax. Indeed, when one enters Savanna, the sign welcoming visitors declares Savanna a 'Sportsman Paradise.' Creation of a refuge at the depot would put an exclamation point to this statement while preserving the rare elements at the depot and providing significant economic benefits to the community," Boyd wrote.

"There's a major land grab going on right now, right here in our backyard. I can't believe that Carroll County residents aren't loudly protesting," Jon Whitney, publisher of the Carroll County Review, explained as the battle between the LRA and Wildlife Service rose to a fever pitch. Whitney asked the USFWS to stop its "land grab." (9)

The USFWS justified its request for the land because of the migratory waterfowl corridor, active bald eagle and heron-egret populations and "two federal candidate plant species and a number of bird, fish, mammal and plant species currently on the Illinois state endangered and threatened list."

"While I do care deeply about wildlife and its future, it seems quite clear to me that the thrust of the LRA is to find the means to redevelop the Savanna Army Depot to replace the lost dollars to the local economy when the depot shuts down. I don't think that leaves room for USFWS to take over the bulk of the real estate at SAD," Whitney wrote.

When it came to preserving land, Whitney felt portions should be transferred to the IDNR because it was much more committed to allowing public access than the USFWS. "Much of the land the USFWS seeks is area they call 'upland sand' prairie and sand savanna. Certainly, that's true, but don't let that definition fool you. That land was simply poor farm ground until about 1917 when it was sold to the government for the base. Much of that land, in fact most of it, is covered by ammunition storage igloos. This long time usage has allowed a variety of flora and fauna to return to the area," Whitney wrote.

"I think the Fish and Wildlife intimidated us," Sullivan remembered later. He believed that he had been tough during the negotiations with the U.S. Fish and Wildlife. "But now that I think about it, I wasn't tough enough. ... The Fish and Wildlife, I think, lied to us. The Army didn't help us. We were trying to be good neighbors and we got screwed ... I got into negotiations with them and I would walk out of the meetings and everything else. I know if we would have been tougher with them, we could at least got that campground down there (depot sporting club). We should have gotten the (U.S. Coast Guard) boat ramp, too" Sullivan added. (10)

In March, LRA officials met with Edgar to speed up the redevelopment of the depot. As a result, Edgar wrote a letter to the Army asking them to expedite cleanup efforts for a prison site because he included money in his budget for a prison. Interest was high for a Savanna prison because it would employ as many as 1,000 people and create opportunities for local businesses. Competition was fierce for another medium-security prison, and Lawfer lobbied hard in Springfield on behalf of the LRA.

"The governor and other state officials all but said Savanna would be selected. They want it to be part of the overall multiple-use plan," Dahlman assured the LRA board.

A compromise agreement on dividing the land was reached in April between the LRA and USFWS. LRA would control about 3,000 acres and USFWS about 10,000 acres. Sullivan said the agreement would grant the LRA more than one million square feet of buildings as well as providing easement agreements for all roads and railroad access on the property. The USFW agreed to allow hunting and fishing with the availability of other recreational activities in the area under their jurisdiction. During the April meeting, participants involved in dividing the land were continually seen going in and out of secret sessions to discuss new proposals. The property lines changed several times when authorities discussed their new arrangements to the public during the open session.

Whitney wrote on April 3 that the compromise was good for all. "From my vantage point and after imploring people in an earlier column to beware of the USFWS land grab, I think the compromise worked out offers a fantastic opportunity for this area to develop a major site which will boost our economy over the next 20 years," he declared.

As soon as the compromise was reached, Judy Cherry, now a former Hanover resident, expressed dissatisfaction in a letter warning about the dangers of building a prison at the depot. (9) "First and foremost, prison jobs don't necessarily go to local people. Employees of the Department of Corrections are unionized and would be given the option to transfer into the new facility. While I am sure there are many fine, compassionate people within the prison system, I fear there are also many others who are as hardened as the prisoners they guard," Cherry wrote. She also mentioned there was a tendency for gang members and drug users to move into areas where there are prisons.

The compromise "is a great victory for conservation in Illinois and is a good example of people pulling together and working hard," concluded Boyd. (11)

An impact statement prepared by Economics Research Associates, Economic Development Services, Sasaki Associates and Mid-State Associates for the LRA said a 1,500-cell medium-security prison would require 450 employees, 75 percent of whom would probably live in either Jo

Daviess or Carroll counties. (12)

About 100 of the jobs could be filled by workers outside the area and would likely add 254 people to the local population with their spouses and children. Wages each year would be about $11.6 million. Businesses in two counties could expect a one-time start-up sale of $450,000 in goods and services to the prison, along with about $663,000 in sales each year.

The direct impacts would lead to eight to ten new jobs in service businesses and 35 to 40 jobs in retail businesses. New housing starts would also result from the influx of employees and additional activity, the report said. In addition, the construction would create jobs and income.

The further rippling of this activity through the economy could be expected to generate another $2.5 million to $4 million per year and 15 to 25 jobs.

The higher pay at a state correctional facility would probably drive county wages up somewhat. County and municipal governments could stand to gain from sales taxes and property taxes generated by prison procurement and additional home construction. The additional annual sales tax was estimated in the report at $43,787 and residential property tax at $284,916.

Consultants concluded the median age in the two-county area was high, 39 years, compared to the national median age of 34 years. The figure was expected to rise to 41.5 years in 2000, which showed a general aging of the population. Current estimated age distributions for 1995 showed 28 percent of the population were 20 years of age or under and 22 percent at 60 years and over.

"The increase in vacation and retirement housing will support this trend. For a number of reasons, this trend should be viewed as troubling. It affects the labor force, schools, retailing and social costs. The depot strategy should, perhaps, attempt to counterbalance this characteristic," the report said.

In Carroll County, per capita income grew over 13 percent between 1980 and 1990, but still ranked below the state average (75th of 102 counties) in 1994. As for education, 43 percent of the population in the two counties graduated from high school, eleven percent graduated from college and five percent of people 18 years of age or older were enrolled in college. The percentage of high school graduates in the study area compared favorably to the number for the U.S. as a whole, but the area showed considerably lower college enrollment.

Between 1980 and 1991, the study found the labor force declined almost 16 percent in Carroll County. The average pay for the two-county area was between $15,000 and $20,000 which is considerably less than the more than $28,000 per year in metropolitan areas. "This labor cost is favorable for business attraction, but it could also reflect a lower skills level," the study concluded.

Besides the 421 jobs lost with the depot closing, the study found schools could lose 158 children or seven percent of the school population. The closing could mean a loss of $912,550 in property taxes each year. Since about 80 percent of the property taxes are devoted to schools, the depot closing would mean an annual loss of about $730,040 in the two counties.

The study also examined the four depot sites proposed for a prison. Site No. 1 in the lower post area was being used as pasture for grazing cattle. It had three tanks containing asbestos, and was in the howitzer firing range during the 1920s. Site No. 2 had 14 storage warehouses and it, too, was used for cattle grazing. Also in the howitzer range, was a guardhouse septic tank and leach pit and ten power magazines which should be demolished.

Site No. 3, the area near Whitton Gate, had eight explosive magazines which would have to be demolished, two septic tanks and a runoff tank. It had not been used for a firing range and was used to graze cattle. Despite what the environmentalists claimed, the report said Site No. 3, (which was eventually picked by the Department of Corrections and considered the preferred site by the consultants), had "no apparent environmental concerns." Site No. 4, located five miles north of the main entrance, also was used to graze cattle and was heavily wooded in areas. "The site is not obstructed by Hanover bluffs and has potential for exposure to neighboring rural residential land uses," the consultants concluded.

It would cost $2.4 million to construct a sanitary sewer system, water mains and storm sewer for the prison. The prison would take 30 months to construct and cost $67.4 million. It would have a total operating budget of $25.5 million a year.

The report concluded that crime committed by escaped prisoners in Michigan and Florida had been minimal and non-violent. It also showed an average rate of three to four visitors per inmate per month. Police in Illinois communities with state prisons interviewed by the consultants felt that visitors carried in some contraband but that prison officials had it under control. Police said, too, that some additional criminal activity on the part of inmates' relatives living in the area might occur, but that these people and their activities were well-known and under control

Researchers also found that a prison in a community did not cause crime to rise, but that residents believed their neighborhood was less safe than a comparable community without a prison. They also determined that residents in a community with a prison called the police more than residents of other communities, even though they experienced less crime. About one third of the residents in the community with a prison felt more police officers were needed. Forty percent believed "undesirables" were attracted to the area, even though the average inmates received only four visits per year.

Almost 32 percent of residents surveyed stated that prisons were the feature they most disliked about their area and 4.6 percent considered moving. Nearly twice as many people owned watchdogs in the community with the prison, and they were more likely to own a firearm for self-protection, the study found.

In Dixon, real estate agents agreed all types of property values had risen since the prison was built. However, they did not attribute the rise in values to the prison, but merely felt the normal appreciation in values had not been disturbed. It did not seem as if the prison affected property values at all. Newer, upscale homes were being built close to the Dixon prison, probably because it is in a desirable area, at the edge of town and near the country club. One real estate agent said that if the prison was to have a negative effect on property values, the country club was canceling out that effect.

Lee County Circuit Clerk Jean Hammil agreed with prison opponents who claimed Jo Daviess County would be strapped with increased court costs with a prison. (13) Hammil had "tons more work." The inmates file a variety of cases through the county court system, such as assault, nuisance, divorces and appeals. Inmates sometimes sue the parole boards or warden when a judgment doesn't go their way.

Although the appeals are heard in the same county as the crime occurred, the county where the prison is located has to do the paperwork. Hammil estimated that the workload has increased 30 percent as a result of the prison. This is the equivalent of one full-time employee.

Lee County Judge Thomas Magdich told the Galena Gazette he saw nearly 20 percent more cases because of inmates. He was on a statewide committee of judges to discuss the staggering costs that counties absorb because most inmates cannot pay the filing fees. Magdich said a large number of inmates have an indeterminate term, for example: 100 to 150 years. This means that their case must go before the prison review board which ultimately decides when the prisoner is released.

Economics Research Associates determined when prison populations are old and inmates convicted under indeterminate sentencing laws, the burden on the circuit courts tended to be heavier. Prison populations with young convicts who had definite sentences created much less work for the circuit courts.

Those who remained with indeterminate terms in the Dixon state prison were long-term inmates such as murderers or cop killers. Magdich said that there were more petty offenses from people who visited the inmates, crimes such as shoplifting and leaving gas stations without paying. "I can understand why people would want a prison in their community. They just have to realize that it is a mixed bag. There is both good and bad that go with it," Magdich added.

Joan and Jerry Grippo of Hanover expressed their concerns in a letter to Lawfer writing that a medium security prison would result in violent offenders and Class X drug offenders living near them. "Prisons squeeze local budgets and infrastructures costing them money due to constant maintaining of roads, sewers, water, electric, police and school systems. Moreover, prisons tie up court systems due to the many appeals filed by inmates while in prison which must be handled by local courts. Finally, released prisoners often locate in the area where they are released, ending up in local court systems for parole violations and re-offenses, draining local courts."

According to information they gained from Karl R. Becker, IDOC deputy director, the Grippo's learned that three of four jobs created by the prison would be filled through transfers of current prison employees. New residents drawn by prison jobs would actually diminish the number of jobs available to local residents.

"Don't just look for a quick fix to the loss of jobs in the area. Look at the long range future of this area which tourism is the major source. Be aware that congressmen from other communities who have pushed prisons in their area, have paid dearly for it the next election when the citizens became aware of the reality of what a prison did to their communities. They were not re-elected," the Grippo's warned.

When the LRA board voted 8 to 2 to apply for an adult medium security prison and a juvenile prison, Gene Gray of rural Hanover who was opposed to the Whitton site said it should be put on the county line. "The infrastructure cost is prohibitive. That's our tax money. As a private citizen, I'm saying let's spend it wisely. The Hanover Bluffs are second only to the Mississippi River as a natural resource," Gray added. "The prison should be sited near Primm's Pond, which will become one of the most used tourist attractions at the depot."

Anti-prison groups
turn up the heat

Chapter 6

As support started to build again for a prison at the depot, a group began recruiting members to oppose the prison. Led by Judy and Clarence "Bud" Cherry of Hanover, the group passed out materials at a June 7, 1996 meeting in Hanover. (1) The Cherry's called the meeting after Edgar raised the possibility of building two more prisons. The possibility had the LRA scrambling to make sure that one of four sites was at the depot. Cherry said that the site picked by the LRA was near where Crazy Hollow Road meets Whitton Road, an area where she lived. Cherry felt there must be a better site and she reiterated that a prison would bring drugs and gangs to the area. "The question is do we want to risk the kind of things that would happen? I don't."

In July as efforts to secure a prison at the depot continued, letters of opposition from environmentalists began to appear in area newspapers. The depot needed grant money to keep it running, and Lawfer started efforts to create an enterprise zone at the depot. In August the Northwest Illinois Prairie Enthusiasts asked for a nursery at the depot.

Rachuy, a Stockton resident who was president of the Northwest Illinois Prairie Enthusiasts, wrote Lawfer stating that his group was opposed to locating a prison or boot camp on any portion of depot land.

"We are writing this letter to urge you to support the proposal of the U.S. Fish and Wildlife Service to transfer the environmentally significant portions of the former Savanna Army Depot to the Upper Mississippi Fish and Wildlife Refuge. It seems that this proposal represents a once-in-a-lifetime opportunity to preserve a piece of Illinois' natural heritage. It is our understanding that the Savanna Army Depot contains one of the largest remaining prairies in Illinois, if not the nation. The size of the property requested by the Fish and Wildlife Service and the many endangered or threatened species living in this area combine to make this a rare and unique opportunity." According to Rachuy, converting this area into a wildlife refuge would seem to make good economic sense as well.

"We are all aware of the economic growth in the area that has been spurred by tourists, most of whom come here because of our unique natural beauty. Increasing the recreational resources of the region will surely enhance that economic growth," Rachuy wrote. And in another letter on Northwest Illinois Prairie Enthusiasts stationery dated Jan. 27, 1996, Rachuy called for preservation of the prairies, savannas and wetlands. He said the LRA had proposed to the Army that the entire base be opened to economic development. (2)

Sand prairies are the most common prairie type remaining in Illinois, concluded Randy Nyboer, regional heritage administrator with the IDNR, in a letter to Wendy Brown, natural resource scientist with Tetra Tech Inc. of Fairfax, Va. Brown was preparing an environmental impact statement for BRAC 95 Disposal and reuse of the depot.

"The sand prairies and savannas found at SAD, although disturbed, are extremely important in conserving the biological diversity of our state. The combination of large, contiguous acreages of prairie and savanna habitat, high concentrations of endangered resources and health populations of grassland wildlife, especially birds, provide an opportunity for applying conservation biology that is of a mid-continental significance. It is this type of opportunity that prevents 'ecological train wrecks' the federal government is trying to avoid," Nyboer wrote. From an Illinois standpoint, Nyboer felt the ecological significance was paramount.

"The Mississippi River stretches 385 miles along the state's western border, Nyboer continued. "That portion the depot occupies is only the state's western border. That portion the depot occupies is the only remaining ecological continuum of floodplain, forest, prairie and adjacent upland forest remaining in the state."

Nyboer cited the following endangered, threatened and rare species found at the depot in his report to Brown.

Species listed as sensitive included the American bald eagle which is federally threatened and state endangered and known to nest in the bottomlands along the Mississippi River. As of Aug. 28, 1995, two active nests and one inactive nest had been identified on the depot. Large numbers of bald eagles also occupy the depot's bottomlands during the winter months, where they feed below Lock and Dam 12, perch along the river shore and roost in the hardwood forest.

Four federally endangered animal species, a mammal and three invertebrates, had been identified both on and in close proximity to the depot, though the animals themselves had not recently been detected. The Indiana bat, Higgins' eye pearly mussel, Iowa Pleistocene snail and Karner blue butterfly were included in the list.

Two state-listed mammals, the river otter and bobcat, have been observed on the depot. The river otter, a state endangered species, has made several appearances at the depot. Sightings have occurred in Pool 12 of the Mississippi River, on the Lock and Dam 12 access road, along Crooked Slough and near the base commander's house. Fewer sightings have been made of the bobcat, a state threatened species, though a bobcat was found dead within a mile of the installation and bobcat tracks were seen in the snow near the north heron rookery.

At least 12 state-listed birds have been found at the depot. These birds and their status include the following: American bittern, state endangered; brown creeper, state threatened; double-crested cormorant, state threatened; great egret, state threatened; loggerhead shrike, state threatened; long-eared owl, state endangered; Northern harrier, state endangered; osprey, state endangered; pied-billed grebe, state endangered; red-shouldered hawk, state endangered; upland sandpiper, state endangered, and state threatened. The red-shouldered hawk has been observed nesting in the bottomlands.

The western hognose snake is the only state endangered

reptile known to inhabit the installation. Two endangered fish, the pallid shiner and western sand darter, have been identified in close proximity to the depot. Both fish were identified in side channels of the Mississippi within depot boundaries.

At least nine state-endangered and three state-threatened plant species have been observed on the depot. The state-endangered plants include bearded wheat grass, shaved sedge, redroot, false heather, hairy umbrella-wort, fragile prickly pear, clustered broomrape, James' clammyweed, and meadow horsetail. The state-threatened species are kitten tails, Gray's umbrella sedge and blue sage.

The state-listed species found at the depot are protected under the Illinois Endangered Species Protection Act, designed to promote conservatism of Illinois threatened and endangered species. The act encourages state and local agencies to consult the IDNR before carrying out actions likely to jeopardize their existence.

During vegetation surveys by the IDNR in August of 1996, three additional plant species, purple rock cress, blue grama grass and whitlow grass, were discovered in the uplands portion of the depot. "Until further investigation, these three plants are considered sensitive due to the fact that they had not been previously identified in the state. At this time, they are designated as 'state record,' meaning that no formal protection has yet been recommended," Nyboer explained.

Ed Britton, Savanna District director of the U.S. Fish and Wildlife Service Upper Mississippi River National Wildlife and Fish Refuge, and John Alesandrini, a conservation specialist with the Illinois Nature Preserves Commission, complimented the Corps for including extensive documentation of the biological resources and threatened and endangered species on the base. But Alesandrini asked the Army to specify how it will require future owners of property to protect the depot's environment, especially the natural areas.

Nancy Hamill Winter, chairwoman of the Nature Conservancy in Illinois from Stockton, wrote Lawfer expressing her grave concern against a prison at the depot. "Surely there are many other more positive things to do with those acres which would have a much more beneficial effect on the region - particularly those related to recreation, tourism and nature education. To date, the LRA has been extremely close-minded and secretive about negotiations on this matter. The negative consequences of installing a correctional facility (such as drugs, associated riff-raff, frightening aura, et cetera) surely make such a project unwise, particularly adjacent to a 9,000 acre natural area which could be a great boon to the area," implored Winter, who owned several farms.

Marion Siedenburg of Savanna favored a prison in her letter. "I am a retired teacher in this area ... A prison would help replace the well-paying jobs being lost in the closing of the depot. The Savanna-Hanover area badly needs sound head-of-household salaries," Siedenburg wrote.

"This area of Illinois has been beset with downsizing of railroads, industry moving out of the Jo Daviess-Carroll County area and general loss of jobs in this area. A prison would replace some of those lost jobs and bring the average wage paid in the area to a head of household pay scale," wrote Phyllis Lambert of Savanna.

J.F. Beegle of Hanover suggested in a letter to Lawfer that the depot would be a perfect site for a research-high tech park, which would create high salaries and steady employment. "It (the depot) has everything: existing substantial buildings, roads, railroad, marine transportation access, electric and phone connections and ample recreation area to offer tenants or new owners and their employees. We have the man power and our colleges could supply the research department."

"The historic town of Galena, which has been well advertised nationally, and the DeSoto House or Eagle Ridge are here to accommodate prospects while they decide whether to locate at the Savanna Army Depot. We can have more jobs in our area with private industry paying taxes rather than a prison supported by taxes. Don't short change our area when a far greater potential is available on this land known as the Savanna Army Depot," Beegle wrote.

More than 150 people attending a hearing on the prison Aug. 14 heard opinions about the number of jobs it would create.

"It is important that jobs be created in order to retain households, their related earnings and the local tax base, all of which keep the economy viable. The best agenda for this region is building a correctional facility," said Tom Messer, president of the Savanna Chamber of Commerce. (3)

Funeral home director Mike Jones viewed a prison as an opportunity to keep his children in the area. "It affords jobs that people can raise a family on, that they can pay taxes and benefits they can retire on," Jones commented.

Harold Buck of Savanna talked with five friends in five different towns with prisons. "They said it was the best thing that ever happened to their communities,'" Buck explained. (4)

While many at the meeting spoke positively, there was opposition to the prison. Judy Cherry, who formed the Depot Development Coalition, quoted from Chicago and Freeport newspaper articles citing prison guard misconduct and gang members being hired as guards. "If the LRA were inviting a major business that would spew pollution into our atmosphere, you would all be standing here saying no. But they are inviting a corporation that would spew gangs, crime and drugs into our community - I say no," Cherry declared.

Bridget Strum, now a former Hanover resident, felt Hanover and Jo Daviess County residents couldn't afford to have a prison. "We're not in a position to pay for roads to a prison, additional police in Hanover, more teachers or worse, more schools, because of the influx of workers imported to work at the prison," Strum rationalized.

In another area, Haring wrote IDNR Director Brent Manning saying that he strongly supported the efforts of Northwest Illinois Prairie Enthusiasts who are trying to obtain an IDNR nursery on depot property. (5) Haring agreed with Rachuy, the Prairie Enthusiasts president, that such a nursery at the depot would allow the IDNR to better serve the people of northern Illinois, create jobs and allow the IDNR to better manage the many listed plants at the depot. "We strongly believe that such a nursery would also be a great complimentary development for other reuse opportunities LRA is pursuing," Haring concluded.

During an LRA meeting in Galena on Aug. 21, many in the audience appeared determined to turn the board meeting into a second forum on siting a prison. (6) John Sturm of Hanover asked the board to explore all the opportunities available for the depot "as thoroughly as you have explored the prison." He urged the board not to commit to a prison before exploring other alternatives. "I would like to request the LRA to consider a two-year moratorium to allow time to develop and recruit other job possibilities," Sturm proclaimed.

Friends of the Depot, a group which organized April 1, 1996, said in a Sept. 4 Prairie Advocate article that it wanted to accomplish economic development at the depot by having an interpretive center and gift shop open near the old stone

Beatty House on the depot. (7) Training a staff to lead special interest tours at the depot was another goal. A third goal of the group was to develop hiking, biking and birding trails with interpretive signs. Locating a prison at the depot is a "non-issue" with the group, mentioned Gene Gray, a member of the Friends' board of directors.

On Sept. 3, consultants released an initial comprehensive plan for the depot saying its closeness to a river made it an ideal spot for an intermodal and distribution center. (8) Other uses suggested by Sasaki Associates included a housing development, golf course, marina, resort hotel and biking-hiking trail.

Consultant Fred Merrill suggested 640 acres be designated mainly along the Burlington Northern-Santa Fe railroad tracks for industry. He also indicated two potential barge terminal sites on an initial conceptual plan map. One of the terminals was in the center of USFWS land, and the second site was in the lower post area which could serve as a recreational marina.

Everything looks good for Site No. 3

Chapter 7

The Depot Development Coalition claimed it had amassed 3,609 signatures against a prison at the depot by November of 1996. (1)

In a letter to the Galena Gazette, Bud Cherry of the Coalition wrote that most of the signers were opposed to a prison. "The LRA still has the ability to deny the DOC's requests and the governor can certainly say no to locating a facility there. Therefore, we cannot stop our protest.

Continue to write letters to Gov. Edgar telling him how you feel," Cherry advised.

The prison sweepstakes continued in earnest during 1997. The Army followed through on a request to make Site 3 ready. The IDNR sent a letter to IDOC saying of the five proposed sites for a prison, Site 3 would have the least impact because it had only two endangered species.

A report stated that a contractor spent nearly 5,000 hours removing 124 pounds of ordnance; related scrap, such as burned ammo containers, hinges, box handles and nails and 5,040 pounds of non-related ordnance scrap from the site. This scrap consisted of old fence and old farm equipment around a homestead area in the northeast corner. Lawfer said the report proved that site 3 was not a pristine area for the majority of scrap was related to the farming operations which took place before the Army took over in 1917. He continued that it also showed the site was not a munitions dump area, with the majority of ordnance scrap coming from the southern boundary.

The site was grazed for many years and had been impacted by road construction and the building of munitions bunkers, an Army report entitled SAD Natural Communities stated. There was a good possibility the vegetation would recover if the cattle grazing was stopped and a burning program established. Forty-one plant species were counted in a 3,017 acres area, which included Site No. 3, by consultants who prepared the report. Dominant plant species included bluegrass, common oak sedge and little bluestem. Other species found were (in part) yarrow, lead plant, rock jasmine, Carolina anemone, cat's foot, pussytoes, tower mustard, sand cress, thyme-leaved sandwort, beach three awn grass, wormwood, white sage, horsetail milkweed, tall green milkweed, silky aster, kitten tails, side-oats grama, hairy grama grass, awnless brome, downy chess, poppy mallow, partridge pea, New Jersey tea, redroot and Geyer's spurge.

Although Edgar's plan for the construction of more prisons was widely known, the Illinois General Assembly had not taken the necessary steps, such as approving a state of Illinois bond authorization for the construction. Approval happened on Feb. 19, 1997, when the legislature passed bond authorization providing $270 million for prisons. Included was the medium security prison at Pinckneyville, selected in 1996. (Savanna had been one of the five finalists when Pinckneyville was chosen). The bonding authority plan allowed for a medium security prison and a juvenile facility.

With Site 3 cleaned of ordnance, the LRA wrote to Allen Grosboll, senior advisor to Edgar, advising that 100 acres of depot land and buildings were available for a prison. "Everything looks good, and the parcel of ground should be ready for transfer this summer," Haring wrote in a Feb. 25, 1997, letter. "The LRA continues to move ahead with its overall reuse and redevelopment efforts. We are very pleased that the state of Illinois and the Department of Corrections have shown such a strong interest in the area."

Haring told the Freeport Journal-Standard in March that depot officials felt confident the state would pick Savanna for a medium-security prison or a youth detention center being proposed by the IDOC. (2) The IDOC threw out those applications from the 1995 search and opened the process to all interested communities. Haring said the LRA would resubmit its application by April 15.

Flack thought Savanna was in a better position than it was in 1995 to land one of the prisons. "One of the problems last time is we didn't have the land completely cleared. Now, we have that land ready," Flack noted.

Haring thought the lessons learned during the last application process - along with Savanna's showing in the finals - should prove invaluable. "It was good exposure and a good experience for all involved. We have had productive conversations at the state level and we've kept our name out there. Both Rep. Lawfer and Sen. Sieben also are supporting this," Haring explained. "The folks here realize the importance of jobs and the payroll and economic developmental opportunities this will bring the area."

During a hot and humid evening on July 17, the Department of Corrections hosted a hearing in Stockton. The need for more jobs offering above average salaries was strongly conveyed by many of the 500 people squeezed into the Stockton High School gymnasium. The hearing was the last of seven held by the IDOC to hear local opinion about a maximum security or juvenile prison in communities. (3)

Haring, when listing opportunities planned by the LRA, asserted that a prison was part of the reuse plan "from Day One." "This parcel of ground is clean, safe and ready for transfer to the Illinois Department of Corrections." He supplied IDOC officials with a long list of infrastructure items located at the site, including the roads, the water and sewer system already at the depot, the availability of electricity and natural gas, as well as the land.

Lawfer told the crowd, the U.S. won the cold war, but Savanna and the area lost because of it. "The state of Illinois needs to do something about this," he added.

Carroll County Clerk Judy Gray read a letter from the County Board Chairman Bill Ritenour, citing the board's support for IDOC to locate one of their two proposed prisons at the depot. Jo Daviess County Board Chairman Judy Gratton also read a letter of support for the prison from her board.

To cheers from the crowd, Savanna resident Bob Knuth announced he believed some people had a "selfish interest" at stake for the "little bit of ground they bought when they came out here from Chicago."

All comments, however, were not pro prison. Bridget C. Sturm of Hanover wrote Lawfer that Jo Daviess County residents would be faced with "additional tax burden" with a prison at the depot. "We are very sensitive to increased taxes for positive platforms as education, but we will not tolerate tax dollars to negative growth as a prison ... many taxpaying citizens in both counties will not let you jam this prison or prisons down our throats."

"This land should be used for the benefit of the citizens of the area to enhance both economic and recreational opportunities. A prison would do neither of these things, only providing a few short-lived jobs during the initial construction phase. The experience of other communities of similar nature and size has been dismal as well as, in some cases, frightening. A prison facility has been proven to have extremely adverse effects on the socio-economic nature of a small community, and would only discourage the enhancement of tourism as well as other kinds of economic development," wrote Hal and June Patinkin.

It was a dark day in August of 1997 for the prison supporters when the depot was passed over again. Gov. Edgar picked Kewanee for the juvenile center and Lawrenceville for a medium-security prison. (4)

"I am tremendously disappointed and very frustrated at this point that this type of economic development opportunity slipped through our fingers," Haring complained.

Sullivan was also disappointed. "I'm upset with the environmentalists. They had known for two years that (Site No. 3 near the Whitton Gate) was the site that we had planned, and then at the last minute they come along and start giving us trouble." The LRA board learned, Sullivan told the Prairie Advocate, that the Illinois Nature Conservancy and local environmentalists had threatened an injunction to stop the project if Savanna was chosen.

Confirming that he was one of the people who threatened legal action, Ingram of Apple River explained, "Anything that is going to destroy that prairie, we are going to be against it. Our position is that area is one of Illinois' most unique nature prairies. We don't want to see it destroyed." While Ingram credited his foundation with threatening to file for an injunction, he said it did not act alone. "There are other organizations out there. We were the ones who got earmarked as the spokesmen. Any lawsuit would be a joint effort between other organizations.".

The Nature Conservancy supported a prison at the depot, but had concerns about one at the Whitton Gate. "I want to be very clear about one thing: we are anxious to work with the LRA to attract a prison that the community supports to the Savanna area," explained Bruce Boyd, executive director of the Nature Conservancy. He said three factors led to the LRA's defeat: the high cost of putting the prison there, its impact on other economic activities, and the nature of the site. Based on a study commissioned by the Conservancy, Boyd believed there was an enormous opportunity to develop ecotourism activities, such as biking, hunting and fishing, at the former depot.

Harry Drucker, president of the Friends of the Depot, insisted his group never thought about suing. The Friends of the Depot lacked the desire and resources to file suit.

Another person delighted the Whitton Gate area wouldn't get the prison was Judy Cherry, spokesperson with the Depot Development Coalition. "We always felt that a prison was inconsistent with Jo Daviess County and all the things the county stands for," she added. The Whitton Gate, also known as site No. 3, was one of the four prison sites referred to in the reuse plan. It was the one preferred by IDOC. If the LRA tried to campaign for a prison there again, Cherry said, "We will continue our opposition."

Both the Northwestern Illinois Prairie Enthusiasts and Friends of the Depot wanted to establish wildlife viewing areas, recreational trails, educational and restorative projects at two locations on the depot, reported the Savanna Times-Journal on Nov. 6, 1997. The Primm's Pond area, a 200-acre site on the eastern edge of the depot, had various threatened and endangered species. (5)

"The proposed park will enhance the housing development planned farther north. It's the sort of facility people would want to buy into and any other type of development would be enhanced by a nature area," Dick Harmet told the Times Journal. A resident of Elizabeth and member of the Prairie Enthusiasts, a group of about 500 in Wisconsin and northwestern Illinois, Hamet announced the group wanted to continue the prairie environment and rename the area Stewardship Park.

The other project, supported by the IDNR, was the Beaty Creek area at the north end of the depot. Led by Drucker, The Friends of the Depot, wanted to build a visitor's and interpretative center complex at the site complete with an information desk, classrooms, educational exhibits, meeting rooms, a gift and book store, food and vending, restrooms, hunting, fishing and camping supplies, canoe and bike rental and educational and interpretative trails.

"We don't want to be a business ourselves. That would be up to local entrepreneurs. We make nature our business," declared Drucker. "Bird watching alone will have a major impact. It is one of the fastest growing recreational activities in the U.S. today and birders spend a lot of money."

For every 100,000 visitors, Drucker estimated $4 million to $5 million would be generated directly into the local economy.

Goals for The Friends of the Depot, organized in 1995, were the creation of a network of trails, establishment of a stop on the Grand Illinois Bike Trail, promotion of recreational boating and camping and preservation of a corner of the world "to enjoy a quiet afternoon of fishing," Drucker told the Times-Journal. The Friends of the Depot's study cited river rafting in Colorado, canoe liveries in Florida and bicycling on the Elroy-Sparta Trail in Wisconsin as nature-oriented activities which had generated significant economic activity in their areas.

In November of 1997, two environmental groups made presentations to the LRA board in support for several environmental projects at the depot. Harmet and Drucker asked the board to support the Primm's Pond and Beaty Creek projects, both supported by the IDNR. They wanted to acquire the land under a public benefit conveyance. The two sites contained about 300 acres of land which would be added to the 9,000 acres assigned to the Federal Wildlife Service and Corps of Engineers.

A short time after picking Kewanee and Lawrenceville for the new prisons, Edgar announced he would not seek a third term. The Copley News Service in a commentary complimented Edgar for his policies which enriched Illinois' natural areas. The article concluded Edgar worked with federal officials and conservationists to preserve 10,000 acres for open space and recreation at the depot. "I don't think anybody has a better conservation record than Jim Edgar," stated Al Grosboll, the governor's senior adviser on education and the environment. "I think Jim Edgar has set a conservation standard by which future governors will be judged."

Claudia Emken, director of government and community relations for the Illinois branch of the Nature Conservancy, remarked, "I hope whoever comes in next will look at Edgar's record and try to continue the good work." While hopes remained that Edgar would pick the depot for yet another prison, the battle between the environmentalists and development supporters continued to rage.

On Oct. 16, 1997, Ingram wrote a letter to the Times-Journal complaining about derogatory remarks made about him and his organization. "I have devoted a great portion of my life to saving the bald eagle and other endangered species. To enlist the help of other people in this effort, I was instrumental in the formation of the Southwestern Wisconsin Audubon Club which evolved into Eagle Valley Environmentalists, the Eagle Foundation and finally the Eagle Nature Foundation. All of these organizations have conducted research, public awareness of endangered species and preservation of the vital habitats these species require for survival," he explained.

Ingram pointed out his organization felt like the late Rodney Dangerfield because they got no respect. "But we do have respect from those people we have faced throughout the courts and the legal system to protect the habitats we have fought so hard to protect. They know of our dedication and determination to save these habitats and species for the enjoyment of future generations. They know that in many instances we have been the only environmental organization that has taken a stand to save these habitats and in, most cases, have taken this stand against government agencies," Ingram wrote.

The foundation believed there was plenty of land already

contaminated on the depot which could be used as a prison. "We do not have to destroy one of the cleanest 100 acres on SAD. This unique prairie cannot be duplicated anywhere else in the state of Illinois. It has the greatest concentration of certain prairie plants and birds that can be found anywhere in the state. Because of this, we feel it should be preserved as intact as possible," affirmed Ingram.

"Beaver" Bob Knuth of Savanna in writing a letter of rebuttal believed he "couldn't sit any longer and stomach the Save the SAD letter" written by Ingram. "According to your letter to the editor, you'd rather put an industry or other development on contaminated land. Gee-whiz, Mr. Ingram, that doesn't sound too healthy to me. You're for the environment and want the Army to clean up the depot, but it's OK to put industry on dirty ground. Many others and myself are confused by both the written and spoken opinions that you've made," Knuth criticized.

Two environmental groups, the Friends of the Depot and the Nature Conservancy of Illinois, hired the consulting firm of Clarion Associates Inc. of Chicago to prepare a report called Maximizing the Economic Benefits of the Expansion of the Upper Mississippi River Wildlife and Fish Refuge of the Savanna Army Depot. (6)

The consultants agreed the depot had a "globally rare sand prairie considered to be most significant because of its scarcity." The prairie covered 45 percent of the depot with the majority being within the munitions bunker and storage building complex. Oak savanna, considered rare, was found at the depot. It occurs in areas of scattered trees where conditions are suitable for continuous grass cover. Typical species in the savanna were black oak, little bluestar and June grass. Other areas of the depot included an upland forest, wetlands and bottomland hardwood forest.

"The importance of each of these natural communities, even in isolation, should not be underestimated. However, the real significance of the Savanna Army Depot is that it is one of only a few sites in North America where these natural communities exist in a large, single unfragmented assemblage," the report determined.

The depot is the home of bobcats, coyotes, red fox, gray fox and white-tailed deer, continued the report. Thirty-eight sensitive species (those listed by the U.S. Fish and Wildlife Service or the Illinois Department of Natural Resources as endangered, threatened or candidates for endangered or threatened status) are found at the depot and include the American bald eagle, the Indiana bat, the river otter and the bobcat.

A prison constructed on site No. 3 would be visible from a large portion of the refuge, the consultants found. "It is also unclear how completely separate access will be provided to this site without significantly limiting access to Primm's Pond and the adjacent housing area. We believe that the prison would have significantly less impact on the refuge, Primm's Pond area and North Shinske Road residential area, if it were moved south into the mixed use area adjacent to the proposed boot camp. This location would have little impact on the other portions of the site, including natural areas out of the Primm's Pond area and would adequately address access and internal circulation issues," the consultants concluded.

Calling the depot a "unique ecological treasure," the consultants said appropriate development of the natural area would provide significant economic benefits to the local economy.

"To ensure that the economic benefits from expansion of the National Wildlife Refuge are maximized, care must be taken in how non-protected areas are developed. Any adjacent industrial development must be low in scale - major smokestack industrial use would likely significantly detract from the refuge. Distribution and modern assembly facilities are likely able to meet this need as long as entrances to the facility are not the same or adjacent to entrances to the refuge. In addition, care must be taken to ensure the protection of the natural water features of the site which would preclude development of a barge terminal," the report noted.

Residential development adjacent to natural areas, when appropriately designed and sited, does not detract from the natural areas. Sleeping Bear Dunes and many other natural areas, particularly those near major metropolitan areas, have residential development proximate to the natural area.

"All commercial and residential development must be planned and phased to concentrate it in areas that will have the least impact on adjacent natural areas. Therefore, development including the prison, should be concentrated in the southern portion of the depot until market demand requires that development move north and west in development-designated zones. If market-supported development opportunities fail to materialize, consideration should be given to converting more of the depot's property to economically beneficial nature-based recreational uses," the consultants recommended.

The report claimed that the region would benefit from tourist and population expansion. "The services provided to tourists will likely require increasing the number of available hotel rooms, increasing restaurant choices, providing new businesses to purchase and rent recreation, fishing and hunting equipment. These businesses are likely to locate throughout the region though many will look for space in close proximity to tourist activity including existing downtowns and near to the depot. We believe that a visitors' or interpretive center will focus attention on the national wildlife refuge and provide a place from which to organize recreational and interpretive activities. The demand created by significantly enlarging the tourist market will ensure that the support services be located in the area surrounding the wildlife refuge such as Savanna, Hanover and Galena."

"The economic benefits of compatible development of the site are significant. We believe that a variety of different uses, well-planned, appropriately scaled and phased, can co-exist on the site and enhance the local economy and the quality of life of the region."

Maximum-security prison derby

Chapter 8

The race for another prison sweepstakes was launched in December of 1997.

State Sen. Sieben told his colleagues the depot was now under consideration for a maximum-security prison. "The public support for this project is tremendous, and the benefits will extend to all of our districts," Sieben wrote in a letter.

Any community that wants to become the new home for

at least 1,200 of the state's murderers, rapists and other violent criminals should dust off their welcome mat, the Quad-City Times reported on Dec. 7, 1997. (1) A new maximum-security prison would be the state's first since Stateville Correctional Center in Joliet was finished in 1925. Each of the other three maximum-security prisons was opened in the late 1800s.

The new prison would be one of the most expensive construction projects ever undertaken by the state, taking three years to complete and creating more than 2,800 construction jobs. Once finished, the prison could have a staff of about 500 guards and other employees. The Times reported the DOC wanted a new prison because the number of prisoners who require maximum-security confinement is increasing. One cause was the truth in sentencing law requiring inmates imprisoned for murder and other violent crimes to serve more time.

IDOC wanted to reduce the population of the four existing maximum-security prisons. The prisons were designed to hold 4,809 inmates, but currently were incarcerating 7,551. Reducing the number of prisoners would mean more could be in cells by themselves, something IDOC considered safe, IDOC spokesman Nic Howell told the Times. A new super maximum-security prison, scheduled to open in the winter of 1997 in Tamms, was to house 500 inmates in near isolation. Howell explained the point of the super max prison was not to decrease crowding, but to remove prisoners who "cause havoc." IDOC would have to have a site larger than 125 acres.

A short time later, James R. Fisher, complex manager for the U.S. Department of the Interior's Upper Mississippi National Wildlife and Fish Refuge wrote a letter to Sullivan saying the Fish and Wildlife Service did not object to using Site 3 for the prison.

Among one of the first items regarding the prison siting was a letter from Paul W. Johnson, deputy assistant secretary of the Army, who informed Edgar that the Army had completed all the major actions for the transfer of the proposed prison parcel. Edgar had commented in earlier prison site decisions that the prison site was not ready and the earlier siting application was denied because the Army had not completed all of the necessary requirements.

One of the prison's more vocal supporters, Tom Robbe of Savanna, wrote a letter to Edgar on Jan. 27, 1998, depicting the prison struggle as a David versus Goliath scenario. (2) "First, Goliath has money and touches base with many big names in distant areas. He comes as the Eagle Nature Foundation Ltd., Nature Conservancy, Friends of the Depot and Depot Development Coalition, which are just a few of Goliath's names. His tentacles reach out to all corners of the state of Illinois but most recently reached from the Chicago land area to rest here upon Jo Daviess/Carroll County."

Through their meetings and speeches, the Chicago interests have spread a false atmosphere of fear about what would happen to the depot, he continued. Through their shrewd word games, "they have systematically stopped this area ... from receiving the jobs that we so badly need, Governor Edgar."

"This has been a ploy and a route that has no basis except to scare and stop what is necessary in an area that needs not only those jobs, but the stabilization for the building of schools for your education program into the year 2000, Governor Edgar. I say (this) straight forward to you governor as a person who has spoken with many young and older people. Yes, we do need and want that prison here at the Savanna Army Depot. If you were to look into the eyes of the responsible working men and women of Jo Daviess/Carroll

County, you would see we need those jobs for the future," Robbe concluded.

In February, as the LRA considered applying for a maximum-security prison, Sharon Cholewinski of the LRA wrote Lawfer asking him if he could get several questions answered. Lawfer obtained the answers in a letter from Sonny Brown. The IDOCspokesman told Lawfer that inmates in an adjacent minimum-security work camp would work as grounds keepers and do minor maintenance jobs. "They are generally non-violent offenders who are within a few months of release from the IDOC and, therefore, escape is not a large risk factor," Brown explained.

In addition, Brown reported inmates spend about an hour a day out of the cells going to and from the dining room, get about an hour and a half of recreation a day and those assigned to a job work up to six hours a day. Maximum security prisoners have fewer visitors than minimum or medium-security inmates. Visitors are allowed only one visit per month with maximum-security inmates, and each prison establishes its own visiting policy. Maximum-security inmates have the longest incarceration and are closely monitored usually under the supervision and observation from gun towers.

Just as area prison backers were gearing up for another prison fight, an article in the Freeport Journal-Standard on March 12, 1998, took some wind out of their sails. (3) Construction of the two prisons at the depot could cost the local counties millions, according to Duane Olivier, Jo Daviess County administrator. Most needed were upgraded or new waste treatment facilities to handle the 300,000 gallons per day the prisons would generate. The prisons also will need 300,000 gallons of fresh water per day, Olivier told the Journal-Standard. It is uncertain whether three active wells and one inactive well at the depot could meet that demand, he said.

MSA Professional Services engineer Mike Gay studied rehabilitating a 45-year-old, 200,000-gallon-per-day water treatment plant at the depot, as well as building a new facility to handle the needs of the prisons and future businesses. Five miles of sewer lines, water mains and roads would be required to meet IDOC demands. Olivier predicted the counties would have to come up with $8.67 million to cover planning, design and construction of water and waste treatment facilities to meet the IDOC specifications. Expanding the water and sewer facilities to support industrial growth would increase the total cost by $3 million to $11.67 million. Several grants would be available to cover most of the cost and plans were to negotiate with IDOC to cover the county's costs.

During a meeting March 2, 1998, the Jo Daviess County Board wanted definite answers on why the county was being asked to pay for infrastructure improvements. Gratton and several other board members said they were led to believe the Army would pay infrastructure costs related to developing the site. "The Army is asking local people to pick up the costs on this. It's the Army's mess down there, and I don't think it should be passed on to local residents," Gratton complained.

Although he supported the prison, board member Larry Lyons of Stockton believed asking the county to help pay for prisons, which is expected to create jobs and economic growth, was "a con game." Also during the meeting, the board rejected Rutherford's reappointment to the LRA board because he opposed the prison plans and had called for creating jobs in other ways. Rutherford had supported ecotourism, calling it "the fastest growing form of tourism."

On March 26, Lawfer issued a press release announcing that legislation helping the depot had passed the Illinois House. It would help spur economic growth and job creation through tax exemptions, license waivers and a series of other incentives. Lawfer explained the multi-pronged approach would allow abandoned Army depots across the state, including Savanna, to qualify to become enterprise zones, an economic development tool created by the state in 1983. "This is one of the best economic development tools the state has to offer. Communities that are losing a major jobs provider like an Army base need help replacing those jobs. Making the depot an enterprise zone would be an important step in bringing new jobs to Savanna," Lawfer noted.

"Not only would this be a great step forward for Savanna, this is needed by communities facing the same situation as us, across the state," Haring declared.

Types of incentives typically in enterprise zones include abatement of property taxes on new improvements, tax credits, income tax deductions, homesteading and shopsteading programs, waivers of business licensing and permit fees, streamlined building code and zoning requirements and special local financing programs and other resources.

Edgar makes prison backers happy

Chapter 9

While attending the legislative session in Springfield on April 2, 1998, Sieben and Lawfer were notified that a Department of Corrections plane would be available for them to use to fly to Savanna. Edgar aide Jim Kaichuk told Lawfer it would be to his advantage to miss part of the session. Since 1992, Lawfer had never missed a session date as representative and he wasn't fond of missing one then. "I removed my voting key so that no one would be voting my switch in my absence.

Sen. Sieben and I had a secretary take us to the airport where we met Sally Brady, Department of Corrections director of personnel, who joined us on the flight to Savanna. We talked about the aging condition of the plane, the pilot's years of flying experience and the scenery. I do not remember any conversation about the prison. Sieben, a licensed pilot, sat in the co-pilot's seat," Lawfer explained.

Edgar flew to Savanna on a different plane and was on the ground when Lawfer and Sieben arrived. While driving north to the depot, the roadside was covered with signs saying "Welcome Governor." At the depot's welcome center, about 300 to 500 people held signs saying "Thank you, Gov. Edgar."

"They evidently had more concrete information than I had," Lawfer recalled. The towns people were excited because they were told Edgar had chosen a 140-acre site at the depot for a $98 million, 1,000-bed maximum-security prison and a 200-bed medium-security prison. School bands played and people cheered. (1)

Among the crowd was, Jim Shaw, a Milledgeville resident and owner of Shaw's Food Pride in Mt. Carroll. He had joined the bus trip sponsored by the Mt. Carroll Chamber of Commerce. Shaw remembered that "it was a beautiful day and I was thrilled beyond belief about the prison announcement. And, I kept thinking, most of the people there favored a prison, but I knew there was opposition to it."

When introducing the governor, Sieben announced, "Yesterday when I got the official word from the governor's office I said 'I hope this is not your idea of an April Fool's joke.' I'm here to tell you today this is no April Fool's joke. This announcement is real. We've tried for this three times. We've come up to the plate and the pitch was maybe a little high or a little low, but this time when the third pitch was thrown, you people here hit a home run."

Standing on a small stage, Edgar told the crowd "I am delighted to announce the state of Illinois will build a maximum-security prison on a site at the Savanna Army Depot." When he finished those words, fifth grade students from Lincoln School in Savanna released red, white and blue balloons.

"My feeling was that Savanna had an ideal site. There was a real need with the closing of the depot and we felt that whatever concerns that had been raised had been answered when I made the final decision last week," Edgar declared.

Economic impact was one of a "whole host of reasons" the site was selected. Edgar explained. "Those other reasons included its somewhat remote location, the opportunity for expansion, and existing facilities that could be used for staff training. One of the major factors is the response of the local community. In no area are you going to have 100 percent support for anything, particularly a prison. It is important to have the vast majority in support and the majority of the people in these two counties were very supportive of the efforts that were undertaken. We don't want to put a prison where people don't want it. We want to make sure that we will find a good reception for these facilities and we feel very confident that this new prison and the jobs it will bring to this area will be well received and will be welcome here in northwest Illinois."

Edgar went on to say that the prison would pump more than $25 million annually into the area economy when completed, and as many as 300 area building trade workers would be employed during the two years of construction. He predicted that the 465 people employed at the new facility would draw almost $20 million in salaries which could turn over several times in the area communities. Most of those hired to work in the new prison will be from northwestern Illinois. "These jobs will offset the 400 jobs the area will lose when the depot ceases operations. The Jo-Carroll Depot Redevelopment Authority worked diligently to secure this new prison as a cornerstone for area economic growth. We heard them in Springfield," Edgar concluded.

Corrections Director Odie Washington told the crowd the new prison is being designed to make it safe for the inmates, employees and community. "Inmates assigned to this new maximum security prison will be under close observation at all times. Inmates will be segregated into small, manageable groups as they move throughout the facility to eat, recreate, go to school or visit family. This lessens the possibility of large disturbances. In addition, the prison design will enable armed tower staff clear sight lines in the event they need to discharge their weapons to dispel a disturbance."

The prison design called for five two-story cell houses and one designated disciplinary segregation for inmates who chose to disobey prison rules and regulations. A program building was to include space for classrooms, a library, an area to visit family and a courtroom with video conferencing equipment to allow judges to conduct legal matters from remote locations. Another building would include dining rooms and kitchens, a gymnasium and laundry. The Savanna prison was to be the first general population maximum-security built in Illinois since the Stateville Correctional Center was completed in 1925.

In a brief speech Lawfer addressed the crowd, "Today is turning out good because of your hard work. This is the beginning of a great day for Carroll and Jo Daviess counties. It is because you did not give up." Lawfer recalled later that he had tears in his eyes when he spoke. "There were people in the audience who had worked to keep the Savanna Army Depot open and lost that effort, but today it appeared that things were falling in place. The enthusiasm was overwhelming."

Sullivan was excited about the announcement and pleased the depot was selected. "With the closing of the depot, the economic impact on the area of both counties was going to be substantial. I think now with the announcement, with the work to begin as soon as it's started, the economic impact won't be as great. We are going to be replacing jobs with jobs right away. The prison is the base tenant that we've needed. The site of the prison is just a small area out of the entire depot."

The LRA will work for a balanced reuse and redevelopment of the depot, promised Haring. "This is not where we are going to stop. We have a reuse and development plan that we'd like to implement and this is just the start. We want to look at housing, other industries, recreational opportunities, education opportunities, ecotourism efforts. There are many, many more options there before us." Haring felt the reaction had been positive. "I have not heard one negative response."

Standing near a model of the proposed prison at the welcome center, Brent Manning, IDNR director, told a Prairie Advocate reporter that all the environmental issues associated with the prison had been settled. The IDOC had studied the site and approved it with the stipulation that the threatened and endangered species on the site either be transplanted or seeds be harvested for planting elsewhere in the depot.

Chief of the Base Realignment Closure Office, Col. R. Gary Dinsick, flew from Washington, D.C., to be on hand for the event because he wanted to show the partnership involved in the process. "I recognize the hardships that were encountered and commend you as you will see success in the future," Dinsick predicted.

Savanna alderman Cheri Canier told a Quad-City Times reporter that she was thrilled with the jobs, but had some concerns. "Currently, our police force is understaffed. I hope the state comes through with additional funding or grants to keep the officers we have and hopefully help us get some new ones."

"Work on the prison construction will put a lot of the local tradesmen to work," Kurt Brunner of Mount Carroll, vice president of the Northwest Illinois Labor Council, told a Savanna Times-Journal reporter. Many of the workers would be local and the influx of out-of-town workers "will be great for the restaurants and camp grounds and other businesses that are directly affected by the prison construction." Another advantage was that a lot of area young people would be able to join the construction project as apprentices, increasing the labor base in northwestern Illinois.

Jo Daviess County Board chairperson Judy Gratton told a Galena Gazette reporter the prison would impact social services, county services, the court system and the sheriff's department. "We've talked with the director (Washington) about getting a dialogue started. We'll have to get all the players together and start talking," Gratton remarked.

Afterward, Edgar delayed his departure from the Savanna airport for about an hour so he could visit some of the antique stores in Savanna. Flying back to Springfield with Edgar, Lawfer had to collect his car there. "Thus I made two trips from Springfield to my home territory in one day. But I felt it was worth it. This is history for northwest Illinois!" declared Lawfer. "The governor has placed a prison in this part of Illinois instead of the traditional area for prisons in southern Illinois."

While traveling back to Springfield, Sieben remembered how happy he was with Edgar's decision. "Having failed to be selected twice before, the third time was truly a charm," he said.

As many area residents were basking in the prison news, environmentalists were not happy. Along with biologists, they claimed the construction might wipe out two plants on the Illinois endangered species list and disturb the rare sand prairie where dozens more endangered and threatened species existed. Other opposition came from people who said a prison, with its lights and razor-wire fences, would be a blemish on Jo Daviess County's beautiful landscape. They believed it would discourage tourism and other clean development at the depot.

The prison issue became the most hotly debated issue in the Jo Daviess-Carroll County area. Local residents made impassioned pleas for creating employment. They wanted their sons and daughters and grandchildren to be able to find decent-paying jobs close to home. Lawfer, who farms with his son, understood their desire. More than a few people cursed the protected species, especially one unfortunate enough to be called James' Clammyweed. The prison backers accused the opponents of caring more about weeds than people.

Mike Jones, who owned a Savanna funeral home, felt the prison would be an opportunity for his children to stay in Savanna. "It affords jobs that people can raise a family on, that they can pay taxes ...," he added. Jones, a 1992 graduate of Savanna High School who later served the county as coroner, said only a handful of his 92 classmates were able to stay in the area because of the job situation.

Bob Knuth, a Savanna resident for 50 years who lived near the depot, told the 150 people attending the hearing that a prison wouldn't be a problem for him. "I've seen the military stock weapons here. We had Italian prisoners of war, more troops than we should have and a bomb explode up here. We lived through all of that. I can't see where there would be any negative part with anything else that would be here that we couldn't handle."

Knuth later wrote a letter to the Savanna Times-Journal saying environmentalists claims the prison site near the Whitton Gate as being pristine are "shaky at best." This acreage as well as thousands more at the depot cannot be considered pristine. The Army bought the depot property 80 years ago from resident farmers. It was farmed ground ... that was worked over for years."

The Army turned that ground, built igloos, warehouses, roads, laid railroad track, he added. "The definition of pristine is original, primitive, unspoiled. I don't think the depot property qualifies. Unique prairie can be duplicated elsewhere. It's happening all over the place, in this state, as well as others. I can't see where the bald eagles have had any problems living on or near the depot. There are several that are nesting in the area and many more come to roost during the non-breeding season, even with the depot actively involved in its daily mission."

The storm clouds of protest began to build and very soon, lightning would strike. The opposition began to group shortly after Edgar's visit to Savanna. A petition with signatures of those who did not want the prison was faxed to Edgar. And, according to the Dubuque Telegraph-Herald on April 9, 1998, a group of Jo Daviess County residents met in Galena to look at legal and political means to fight the prison. (2)

Members of the group disagreed as to whether they could legally fight the prison based on environmental issues, the Telegraph-Herald printed. However, Terrance Ingram of Apple River said that if any federal funds were used in the prison's construction, an environmental impact study must be prepared. Ingram told the group he knew of an attorney who could start the next day to fight the prison on legal fronts. The attorney had successfully battled mining and road projects in other areas of the state.

"We need to hire an attorney who is smart and savvy. Perhaps an attorney could show that the governor's office and the corrections department skirted proper procedures for public input when they made the selection," declared Gene Gray of Hanover.

If there was to be a prison, the opponents agreed the state should put it in Carroll County. Later, LRA board member Rachuy made a motion at a meeting to put the prison in Area H in Carroll County. "Were the prison put in Carroll County, it would face less opposition from Jo Daviess County residents, and it makes sense to place the prison near other areas pegged for industrial and business development, instead of near an area slated for recreational use." The motion, was defeated 6-3.

When the Galena City Council met on April 13, 1998, they listened to the concerns of many citizens who opposed the prison because of its impact on tourism, property taxes and particularly on the quality of life in Jo Daviess County. In voting four to one to oppose the maximum security prison at the depot, Mayor Dick Auman explained that in addition to its citizen's concerns, the council considered in its motion the lack of communication between Edgar's office and local elected officials regarding the decision to place a prison in the county. Jim Baranski, a Galena resident who opposed the prison, believed the village board's vote of non-support "should send a strong message to the county board."

Gov. Jim Edgar (in photo above) told a big crowd at the Savanna Army Depot on April 2, 1998, that he has selected the depot for a prison. Below, Edgar tells the media all the environmental concerns for the prison have been resolved. Next to Edgar is Illinois Department of Corrections director Odie Washington. Illinois Department of Natural Resources director Brent Manning is in the background.

The Savanna High School band (photo above) played before a big crowd at the depot prison announcement. Aerial view of the depot (below) shows in middle where the prison was planned to be built.

Army Environmental Coordinator John Clarke (photo above) stands by the Coast Guard boat ramp and Crooked Slough at the depot. More than 14 miles of rails were torn up from the Lost Mound unit (below) of the Upper Mississippi River National Wildlife and Fish Refuge at the depot.

Headquarters of the former Army Depot is now the offices of three Army employees.

The old stone house was once used for the Underground Railroad.

Lost Mound unit (photo above) of the Upper Mississippi River National Wildlife and Fish Refuge dominates most of the former depot. Aerial view of depot (below) shows lower post in foreground.

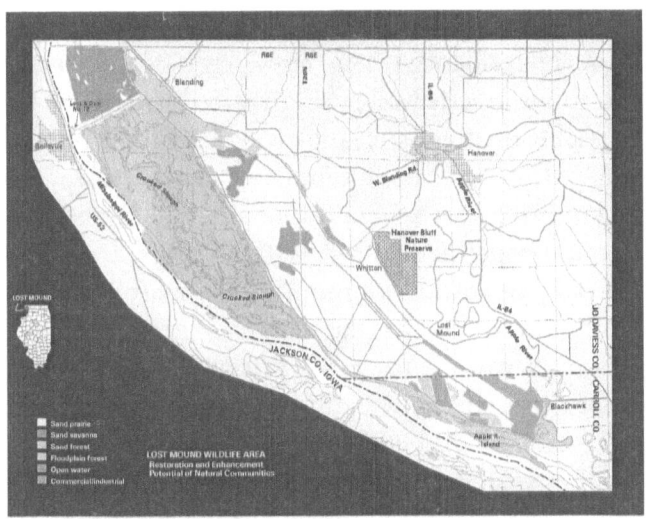

Map (photo above) identifies the sand prairie. Nature is taking over most of the depot as a forest is starting to form along the River Road. Lost Mound unit of the Upper Mississippi River National Wildlife and Fish Refuge headquarters is below.

A few days later, Galena's opposition was shared by many Hanover residents when the Village Board voted three to two to write letters opposing the construction. Among the 50 people in attendance mostly opposing the prison, was the Village Clerk, Susan Fulton. The 25 year resident of Hanover told the Telegraph-Herald that she had mixed feelings. (3) "I can sympathize with how they feel, but we could really use the jobs in the community," she reasoned. Trustee Carl Mutters voted against the motion of non-support, "not necessarily because I favor the prison. I didn't feel it was the proper time because I didn't have enough information to make an informed decision."

"Edgar's April 2 announcement came as a surprise to most Hanover residents and officials, and I think people feel as though this deal got secretly done and that the governor is pushing something down their throats that they had no voice in," explained John Sturm of Hanover.

Many did feel betrayed by Edgar, a lame-duck governor, that he would pick the prairie for a prison. Several days after the announcement, a group of about 80 people gathered in Galena to formalize opposition to the maximum-security prison. The group hoped to use political and legal means to fight the prison. They claimed the state did not receive input from local residents, nor did it adequately assess the potential environment, said a report in the Dubuque Telegraph.

"The governor's office reassured us frequently that it would not locate a prison where there is not overwhelming support. But the Jo Daviess County opinion is being ignored," declared Nancy Winter of Stockton. Winter wrote a letter to Edgar which appeared in local newspapers. (4) "Your sudden appearance at the Savanna Army Depot last week to announce that the state of Illinois intends to build two prisons on that property made a mockery of the whole democratic process. It is a sham. Just because you are a lame duck does not give you the privilege of abusing political power. Autocracy is not acceptable in the American political system."

Winter had asked for the advice of qualified Illinois Department of Natural Resources employees about the location. Her request was ignored. "They (Jo Daviess County residents) have been stabbed in the back. As you know, an alternative potential location, site H, has been repeatedly proposed. Every logical reason favors it. Regretfully, manipulative politics are molding decisions, not sound thinking."

In an April 15, 1998, letter to LRA and elected officials of Jo Daviess County, Winter stated that determined people were out there who would fight to save the endangered species, and litigation could tie up the land for a much longer time than it would take to clean up Site H. "The Army could be pressed to accomplish the required clean up swiftly; there is no question about that. This hassle of lawsuits could be avoided and local jobs could be created sooner if the site were changed," she said.

Lawfer declared later this was the first time he became aware of anyone threatening litigation to stop the prison construction.

Another group claimed they were concerned about a maximum security prison in the area. The group, composed of store owners, property owners and residents, said they were gathering information on the impact in other communities where agriculture and tourism needed to be improved.

In an opinion piece in the Freeport Journal-Standard on April 15, 1998, Carroll County Board Chairman William Ritenour of Milledgeville wrote that he had talked to county chairmen in Livingston and Lee counties, which both have prisons. Those counties were doing a good job of attracting industry and commercial business enterprises. (5) Ritenour also addressed opposition to the prison, disagreeing with statements and saying that the character and the perception of Jo Daviess County would be altered by the prison.

"How many people look up information on perspective tourism centers to see if a prison is located in that county before they travel to the area? I also believe the governor's office did do its homework on the best possible site for this new prison. I firmly believe the Department of Corrections will be a good neighbor for both Jo Daviess County and Carroll County," Ritenour wrote.

Others were not in agreement. The announcement of a prison shocked Linda Howard of Hanover and left her dumbfounded. "How could this happen when the people of Jo Daviess County were so assuredly against the prison? This announcement was a mistake. It makes a joke of our democratic system. We are told that we want a prison when we overwhelmingly do not. You have been listening too close to the LRA which wants jobs at any price. The price could be the destruction of tourism in Jo Daviess County and the quality of life that we enjoy."

Howard, who lived on West Blanding Road located north of the depot, wrote in an April 10, 1998, letter, to Lawfer asking him to withdraw support for the prison. "We have based our economy in Jo Daviess County on small businesses, agriculture and tourism. We do not have high unemployment. East Dubuque announced last week the relocation of two businesses to that community. The residents of Hanover, in a community study, voted overwhelmingly to not have a prison, and I am reading in the press that the residents are in favor of this prison."

In an April 19 report, the Dubuque Telegraph-Herald interviewed officials in other communities with state prisons and found prisons mean more advantages than disadvantages. Many of the officials said the fear of crime is over-rated, and for the most part, local economies have improved because of the prisons. (6)

Carol Hoffman, sales manager for the Southernmost Illinois Tourism Bureau in Ullim, hadn't heard any negative comments or changes in tourism because of the Tamms Correctional Center near the Shawnee National Forest. If anything, the prison, opened in 1995, has brought more visitors because people are curious about it she explained.

The Anamosa State Penitentiary in Anamosa, Iowa, known as the "White Castle of the West," has become more of a tourist attraction than an economic liability, residents have been known to say. The 124-year-old prison is built of rare white dolomite and is on the National Register of Historic Places. The unusual architecture attracts tourists. "If you live here, you don't even know it's there. Even though it's close to downtown, it's a quiet place," remarked John Haldeman, Anamosa's city administrator.

The Western Illinois Correctional Center, located in an industrial park on the edge of Mount Sterling, Ill., does not turn off visitors, stated Larry Dunn, treasurer of the Brown County Development Corp. "I've had more people tell me the industrial park is just beautiful."

The boyhood home of Ronald Reagan and the scenic Rock River, Dixon, Ill., is the site of the 2,100-bed Dixon Correctional Center, located 100 miles west of Chicago. Families of some of the prison's inmates have moved to Dixon. "It was believed that this (inmates' families moving) wouldn't happen. Well, it has. That can have no impact or it can have some impact," remarked former Lee County Sheriff Tim Bivins. The families visiting might not be a problem, but other visitors might be. "There will be a gang influence because it's in the prison. That brings people related to gangs to the community."

Some inmates, upon release, chose to stay in the Dixon area. "People said, 'oh no, they usually go back' (to where they're from). But that hasn't been exactly true," Bivins added. Just as families of inmates shouldn't be stereotyped as troublemakers, neither should former inmates. "Some of the former inmates are quite productive. They're working, going to church. They're rehabilitated. They're leading productive lives. They're contributing members of the community."

Brown County state's attorney Jerry Hooker remarked, "we've never had any people visit the prison who have committed a heinous crime. Have we made some traffic stops and found some (marijuana) as a result of increased traffic from visitors? Yes, we have." As for increased court costs, the addition of a prison had cost Brown County little, he added.

The sheriff of Brown County, Michael Myers believes that friends and relatives visiting inmates at the Western Correctional Center, which opened in 1989, have caused few problems. "We have had maybe a dozen or fewer families move here to be closer to someone in prison. There has not been a major migration of individuals, maybe because this is a rural community and there isn't much to do here. Those who moved to the area have been law-abiding because people would notice if they were not because the county has a population of only 6,000," Myers noted.

Anamosa District Court Judge August Honsell told the Telegraph-Herald that there isn't much of a problem with additional caseload because of the prison. "There are some prison disciplinary proceedings which go to district court, but prisoner litigation is controlled somewhat by the state policy requiring pre-payment of filing fees."

Bivins, however, saw more impact. "The courts will see

more activity. The inmates start filing motions to change their names, get married, get divorced or they'll sue. It's just a ploy to get out of the cell and get into court. It does add a burden to the court system."

As for employment, the super-max and work-camp prison at Tamms employ about 450 people, but "very few" of them are local residents, explained Carl Hileman, former president of the Tamms Chamber of Commerce. Many are transfers from other prisons. The Mount Sterling prison recruited about 20 percent of its employees from the local population. "Twenty percent of 425 is 80-some jobs," Dunn told the Telegraph-Herald.

During prison construction, communities have found that some local contractors and construction workers have been hired. However, in small communities, few have experience building prisons. Although most major contractors do come from outside the area, a positive economic effect can still result. Walter Pang, president of the Tamms Village Board, remarked, "They (the contractors) buy groceries, they get gas, and they eat in our restaurants."

No one contacted by the Telegraph-Herald cited any evidence indicating prisons discourage businesses from developing. In Jacksonville, Ill., for instance, an industrial park was established across the road from the prison and a housing development sprang up within a half mile. "What the Western Illinois Correctional Center has given us is stability. It has given people the confidence to build their businesses, which they would not have had without 450 good-paying, clean jobs," Dunn declared.

Sue Fogarty, owner of Jones County Realty in Anamosa, could not imagine what the town would do without the prison. "It's our biggest employer. Probably 20 percent of couples who walk through my door looking to buy a house have one or the other working for the prison."

Geri Stevens, an agent for Stevens Realtors in Dixon, believes that the prison has not hurt property values. "I haven't had any trouble selling homes on the street that runs right behind it. The property certainly hasn't depreciated."

Housing values have risen, not fallen, in Taylorville. Bob Craggs, a real estate broker appraiser, declared, "The real estate market has been fairly good ever since the prison opened." In Jo Daviess County, the Telegraph-Herald reported the housing market was tight, especially in low to modest-priced homes.

Prison jobs versus preserving the prairie

Chapter 10

In the spring of 1998, as the environmentalists planned their next moves, prison supporters took pen in hand.

"Treating Savanna prison and golf course (at Eagle Ridge) differently is totally wrong," headlined a letter in the Clinton, Iowa Herald by Savanna Army Depot employee Thomas Robbe. (1) Robbe wondered where all the environmentalists were when the Galena Territory golf course was proposed.

"From what I understand, it takes some 65 to 90 acres

- or more than 7,200 yards - to make an 10-hole signature golf course. Hopefully, this area was checked by the Illinois Department of Natural Resources and other environmental groups. To protect those endangered and threatened species, such as squirrels, deer, foxes and rabbits; to all the oak, hickory, trees, the insects and Mother Nature's other wonderful works that may have been moved by no fault of their own - for the sake of a golf course we clear acres of land. How nice," Robbe wrote.

For the environment, Robbe believed that it seemed like people would do anything. But for 100 acres of sand far away from people and land that had never been on the Illinois tax rolls that might provide a decent living for the population, "the world comes to a standstill. I think there is something totally wrong when on one side of the environmental fence, there is one rule and on the other side a different one. Why can so few people stop something that would benefit so many for the future in both counties on superficial environmental issues, yet stand by while entertainment growth rolls over the hills?" Robbe asked.

Based on natural resources, Fish and Wildlife Service District Manager Britton reported Site No. 3 was an acceptable location for a prison, provided that concerns by the Illinois Department of Natural Resources were addressed for state-listed plant species. "A prison would provide long-term economic gains for the community and would provide an anchor tenant for infrastructure improvements and future economic expansion efforts at the depot," Britton wrote Jo Daviess County Board chairman Gratton on April 20.

Anti-prison activists stepped up their attacks later in April. At an April 8 meeting at the Eldorado restaurant in Galena attended by about 80 people, several in the audience asked the Jo Daviess County Board to stand against the prison. Others also discussed possible legal action, letter writing campaigns and petitions.

Later in April, the Jo Daviess County Board gave support for the prison. 650 people had packed the River Ridge High School gym north of Hanover. The vote came after about two hours, during which many in attendance had expressed views on both sides of the issues.

IDOC Director Odie Washington answered questions at the meeting. (2) Comments during the evening dealt with fears of increased crime, drug and gang activity and their effect on the communities and tourism. However, others said the prison would be a source of economic opportunity which would allow the young to return to the area.

"The information that we have been provided with is quite overwhelming. The prison is going to be safe. There have been a lot of red herrings put out about drugs coming to the county. I hate to tell you this, they're already here. Bringing a prison isn't going to change that," proclaimed John Creighton.

Jim Baranski of Galena proposed the prison be built in Carroll County. "Part of the problem that we have here is that we have two very different counties. Both have different characteristics, both certainly have different goals. The problem is that the people in Carroll County ... very much want this prison, particularly the people in Savanna. The problem is that the people of Carroll County want a prison very badly, but they want to have it in Jo Daviess County. I would never deny the right of the people of Carroll County to have a prison in Carroll County. However, I would never recognize their right to choose and put a prison in Jo Daviess County."

Cathy Brunner of Mount Carroll, Carroll County's zoning officer, told the crowd that her parents live within two miles of the access road (for the prison). "My friends and family live there. We're not afraid of it being there; I'm not afraid of it being there. I believe there is room for all of the things that everyone has been talking about. I feel that it can support people in this community. People my age and people coming out of high school will be able to stay in this area because of the offshoot jobs that will come from it and the construction trades, as well as tourism. I don't feel that tourism jobs are the only kind of jobs that this area will support. I believe agriculture, prison and tourism will work in this county if we all work together."

Disagreeing, Holly Marks of Elizabeth declared that she didn't want any of her children to work at the prison. "I don't want any of my children around gang influences. I don't want any of my children around drug traffic that I know will come. We're just opening the door and saying we're here because we want the money. I'm kind of curious, people talk a lot about money. Is that going to be the end all value .... (in) this whole decision?"

Sam Haas of East Dubuque felt that young people need good paying jobs so that they can stay in the area. "We need jobs, and I think this prison would be just the start of many businesses coming in this area. I've heard a lot of people talk about tourism. One of our places to visit is Door County, Wis., and every time we drive up to Door County, we pass the prison in the Green Bay area, and that doesn't deter myself from going to Door County."

John Sturm of Hanover thought the only voices that should be heard were people who pay taxes in the county where the prison is going to be built. Believing the public process had been ignored, Sturm surveyed many parents and younger people. "They're absolutely insulted by Gov. Edgar thinking they are waiting to get out of high school to go to work in a prison facility."

Whitney, publisher of the Carroll County Review, found it unusual that so many in Jo Daviess and Carroll Counties had failed to grasp the economic significance of the best "opportunity to occur in this area in the last 50 years. Almost 31 years ago, my wife and I chose to leave college when I was within one semester of securing my college degree and follow my dream of owning my hometown newspaper. At that time I remember being sure that the area could do nothing but continue to grow and prosper, particularly because of the influence the Mississippi River would have on the tourism market and also because of the potential that relatively cheap land here offered industry, making some major expansions throughout the Midwest."

"In that 30 years I've watched as we have seen the number of railroad workers based out of this area dwindle to little more than a handful. The numbers of persons employed at the Savanna Army Depot continued to drop year after year. Shimer College closed its doors in Mount Carroll.

"Several other factories, particularly in Savanna, either moved out or closed. The farming industry continued its fast-paced evolutionary change and the numbers of available jobs on the 'family' farm fell beneath the wheels of the ever larger tractors which rolled over ever larger acreages."

Whitney believed that all of those events contributed to the tremendous decline in the number and variety of stores on Main Streets within the county's seven communities. The population of Carroll county dropped one-third, and "we continued the trend of exporting our brightest, youngest and most talented to larger cities which offered a better chance to fulfill their vision of the American dream," noted Whitney.

When the government decided to close down the Savanna Army Depot (SAD), the decision was viewed by most as another blow to the community. "Yet, that decision may be the catalyst to bring an era of economic prosperity to this area which has not been seen since the days of the lead mines in Galena and lumber mills in Clinton," Whitney wrote. He was upset with those who told "outright fabrications" about the prison site. "First, there was no secrecy about the desire of the LRA to secure a prison at SAD, nor was there any secrecy about the state's desire to locate one on a piece of property they selected almost two years ago. Some persons are upset because there were no additional public hearings. I'm not. How many times do we have to sit through the same type of meeting, hearing the same pros and cons from the same persons?" Whitney asked.

In a letter to the Galena Gazette, Roland Unangst of Hanover supported the prison on behalf of all the young people. The prison would provide a good place to work, just as the depot provided a good place for him for 32 years. "And to those of you out there who place tourism, grass, birds, bike trails and the likes ahead of jobs, I feel sorry for you. Just how you can look yourself in the eyes while you shave or put on makeup is beyond me. No, the Hanover Ambulance, the Jo Daviess County Court Room, local medical personnel and local hospitals will not have to serve 'the Max' prison. You see," Unangst summarized, "the prison comes with all of the aforementioned as standard equipment."

President and founder of Friends of the Depot, Harry Drucker was one of the more vocal opponents to the prison. Drucker was a farmer and operator of a commercial cattle feed lot in Jo Daviess County, and served as treasurer on the board of trustees of the Illinois Chapter of the Nature Conservancy. Addressing an Illinois Nature Preserves Commission meeting May 5, 1998, Drucker emphasized that Site No. 3 contained threatened and endangered species and was an extremely rare natural community. (3) "A 140-acre prairie is very important prairie anywhere in Illinois, but this prairie is even more important because it is an integral part of a much larger assembly of many different types of natural communities all knit together in one unbroken continuum."

Drucker believed Site No. 3 was part of a larger prairie, which included 10,000 acres and more than 40 threatened and endangered species. "The depot's upland contains the largest continuous tract of sand prairie, and savanna remaining in the state, and the depot is further identified as one of the largest remaining natural grasslands in the Midwest. Site No. 3 is right in the middle of these uniquely, unfragmented natural areas. What possibly could be the compelling justification for building anything on this site. Is it because the state will get the land free from the federal government?" Drucker inquired. He explained that alternate, superior sites could be made ready and were available from the government at no cost to the state.

He wondered about an alternate site, called Area H, although not ready for immediate construction, it could provide everything the Department of Corrections would need. It was larger than Site No. 3, with 40 acres of it covered with old World War I era warehouse buildings all in such disrepair that they needed to be torn down. The prison site occupied 140 acres, only 35 were required for the prison building. "We should follow the example of the old Joliet Army Arsenal where many obsolete buildings just like the old warehouses at Area H ... were torn down," Drucker recommended. "Why not raze the buildings and construct the prison in Area H?"

"This alternative (Area H) provides the Department of Corrections with the land it needs while avoiding an extremely negative environmental impact. Most people are unaware of the unique ecological treasures on and around the

Savanna Army Depot." Drucker supposed this was understandable when you consider the Department of Corrections could have overlooked the significance of the prairie at Site No. 3. The U.S. Fish and Wildlife Service originally wanted all of the area except the developed end of the depot, "but it looked politically bad, like it was a land grab," Drucker told the commission.

"Fortunately, there is still time to act, but time is of the essence because the Department of Corrections has indicated that it intends to break ground as soon as possible ... It is the fervent hope and desire of all who wish to preserve our unique and irreplaceable natural heritage that this commission draft a resolution or letter to Gov. Edgar strongly urging him to recommend construction of the prison on an alternate site of the depot or Brownsfield or already disturbed sites such as Area H, but not on Site No. 3," Drucker pleaded.

Carolyn Grosboll of the Illinois Nature Preserves Commission, whose husband Alan worked with the IDNR as an aide to Edgar, told the commission that she had researched the issue and thought the LRA rather than the IDOC had recommended Site No. 3. Grosboll did not know what authority the Department of Corrections had in building a prison in one area or another. These were questions which Grosboll felt needed to be asked of the IDOC.

But before that happened, the commission's meeting concluded when it recommended letters be written to Edgar and the IDOC asking them to reexamine their decision to build the prison on Site No. 3. Recognizing the conservation and preservation values at stake, an alternative site or sites should be considered, the commission urged.

On May 15, Drucker gave testimony at an Illinois Endangered Species Protection staff meeting. He was introduced as founder and president of the Friends of the Depot, a not-for-profit organization dedicated to promoting economic redevelopment at the depot "in a manner consistent with the preservation of its natural and cultural resources." Drucker identified the Whitton Road prison site as "140 acres of pristine prairie." He also said Site H could be made ready for a prison within 18 to 24 months.

In a May 22 letter from commission chair, Victoria Post Ranney urged Edgar and IDOC officials to reexamine their prison decision. "Site No. 3 contains 140 acres of rare high quality prairie, including species protected under the Illinois Endangered Species Protection Act. The area is adjacent to Hanover Bluff Nature Preserve, a 362-acre area dedicated as a state nature preserve under the Illinois Natural Areas Preservation Act. Further, this area is designated as an Illinois Natural Area Inventory Site. The placement of a correctional facility at this site would destroy protected species and may negatively impact Hanover Bluff Nature Preserve, thus destroying a portion of our rare and important natural heritage," added Ranney. She reviewed the purpose of the Illinois Natural Area Preservation Act which provided that the Nature Preserves Commission keep watch over the protection, management and use of nature preserves. Also included was the commission's responsibility "to promote by advice and other assistance the protection of natural areas in the state which are not dedicated as nature preserves."

Caroline Grosboll, waited until July 31, to notify Lawfer of Drucker's letter and presentation which had lasted more than the allowed three minutes. Caroline was the wife of Allen D. Grosboll, senior assistant to Edgar who worked on special projects including the prison siting. Lawfer had been unaware of the INPC action and the letter to Edgar.

Later, after learning of the letter to Edgar, Lawfer said

that he felt the board had exceeded its statute authority. He explained that the IESP act clearly stated, "the board shall also advise the department on methods of assistance, protection, conservation and management of endangered species and their habitats and on related matters. It does not authorize the board to send letters to the governor."

Despite the uproar, Edgar stood firm on his prison decision. Eric Robinson, an Edgar spokesman, told the Dubuque Telegraph Herald on June 8 "the governor has not changed his position." (4) Unless Edgar picks a new site, the IDOC sees no reason to change its direction either, explained IDOC spokesman Nic Howell.

In its June 9 letter to Edgar, the Illinois Endangered Species Protection Board pointed out that Site No. 3 was known to contain two endangered plant species, shaved sedge (Carex tonsa) and James' Clammyweed. (6) "It is sometimes tempting to think that if it is only a plant, an endangered species can simply be dug up and moved to another site. The board feels that this is only an appropriate strategy when all other alternatives have been exhausted. An endangered plant which is highly specialized in its habitat requirements cannot simply be moved and expected to thrive as it did in its original location," the IESPB wrote. The group felt it would be unfortunate to break up such "a large block of native sand prairie and further fragment this important grassland habitat if there were other suitable areas where the prison might be located."

In response to a letter from Drucker, Fish and Wildlife Service complex manager James R. Fisher reported that his department had raised concern about four prison sites at the depot because they contained state-listed species. "However, in light of the impending economic impacts of the depot closure and my belief that the sites designated as potential prison sites were the only alternatives that would be considered by the Department of Corrections, we agreed not to stand in the way of a prison on one of those sites provided that the state's concerns about the listed species could be addressed to their satisfaction. With that as a background, and in response to a specific request from (LRA board chairman) John Sullivan on Dec. 12, 1997, I sent a letter stating that the service would not object to the use of Site No. 3 for a prison site."

"It was not my intent to imply that Site No. 3 is our first preference for use as a prison site," Fisher explained. "Consistent with the concerns that we have expressed about the presence of the state-listed species, if there is a site in the designated LRA portion of the depot that would serve as a suitable prison site and that would involve less environmental disturbance, we would find that preferable to the use of Site No. 3."

The alternative site offered by the commission was a designated contaminated area. John Clarke, the Army's environmental coordinator at the depot, thought it could take several years to clean it up as a Brownsfield site, (a polluted area for which federal funds are used for cleanup). Had the Army cleaned up the site two years ago, Clarke said it would be nearly ready now. "We've been working for over two years to get the LRA to submit other sites," he explained.

Meanwhile, the IDNR believed an endangered plant at Site No. 3, the James' Clammyweed, could be transplanted to another depot location, but a scientist with the Illinois Natural History Survey said that was not the case. Geoffrey A. Levin, in a memo, wrote his department chief Dr. David Thomas that it was unlikely that small annual plants could survive transplanting. (5)

"Site No. 3 is primarily dry-mesic sand prairie with many blowouts. It is home to a number of rare plant species.

Especially notable is James' Clammyweed, Polanisia jamesii, an annual plant that grows on blowouts at SAD, but currently nowhere else in Illinois. It is listed as a state endangered species. It is not a strong candidate for migration by transplanting because it is exceedingly unlikely that the small annual plants would survive transplanting." Levin continued, "Furthermore, it is dependent on the specialized blowout habitat and would not survive in the typical prairie habitat at SAD. It is unknown whether the prairie vegetation can be artificially manipulated in a way that would enhance James' Clammyweed survival and growth."

Another issue Levin associated with the development of Site No. 3 was the impact on the overall biological integrity of the depot. Because of its location along the eastern perimeter of the depot, close to the Hanover Bluff Nature Preserve, Levin felt the site has particular importance.

"Developing the area would further isolate SAD from Hanover Bluff, significantly increasing the habitat fragmentation disproportionately greater than its limited size would suggest. Other development planned along the eastern edge of SAD would create a highly detrimental cumulative impact on the biological integrity of the area."

Thomas forwarded Levin's memo to Jean Gray, an activist with Friends of the Depot, a group which opposed Site No. 3. Gray told the Dubuque Telegraph-Herald in a June 9 report that the Illinois Nature Preserves Commission had written a powerful letter to Edgar. Edgar had a very good environmental record. Gray urged, "there has got to be some rethinking." Haring responded by telling the Telegraph-Herald in the article that Site No. 3 had been the only site under consideration. "I don't know why the site issue needs to arise again."

Joining the fray were three more environmental groups,

the Sierra Club, the Northwest Illinois Prairie Enthusiasts, and the Illinois Endangered Species Protection Board. The Sierra Club asked its members to contact Edgar by mail or fax to build the prison on land not considered pristine. In a sample letter, the Sierra Club stated that an alternative prison site on developed land would benefit the state, region, environment and wildlife. The alternate site would also provide the opportunity to develop an eco-tourism and recreation base for the region. "Please move the location of this prison to protect rare ecosystems and species and save taxpayers money. I would appreciate any measures you can take to protect these natural areas for our families and our future," ended the sample letter to the governor.

LRA board member Jim Rachuy, in his role as president of the Northwest Illinois Prairie Enthusiasts, wrote letters to Edgar, Sieben and Lawfer advising that a maximum security prison and nature-based tourism "are incompatible opportunities. That is, an improvident siting of a prison and could ruin the prospects for outdoor recreation and tourism. Land use planners routinely work to minimize such incompatibilities and to maximize the benefits of the total development program. From economic development, political, financial and ecological perspectives, Site No. 3 is a poor location for a prison. The pending political and legal challenges to the prison from environmental organizations would quickly disappear with the selection of a Brownsfield site," Rachuy insisted.

In response, Sieben told Rachuy that Site No. 3 was consistent with the consensus plan which was developed with all parties involved for reutilization of the depot. "I also believe that there is ample opportunity for the Prairie Enthusiasts to develop, promote and market nature-based tourism on the nearly 10,000 acres of the depot set aside for these purposes," Sieben wrote.

Another blow to the prison plan came from the late John Husar, a Chicago Tribune outdoor writer, who wrote that the prairie land was too precious to house a prison. (7) Husar called prison officials "shortsighted" to select such a high-quality prairie for a prison. "The prison would be located smack against a 362-acre nature preserve amid some of the lushest, most scenic wild spots beneath the Mississippi River bluffs. Those who have conceived of this atrocity must be asked if they intend to take a stroll beside a gloomy prison. Where, oh where, in the world have they buried their heads?" Husar cried.

Husar's sources within the Department of Natural Resources told him the decision was "a quick fix effort to nail down a prison site, well before the long-term environmental impact could be measured. The chief idea was to satisfy local economic interests that, admittedly, could not care less where the cellblocks are located, just as long as they are close to town." The trouble was, no one looked at the long picture, the source said, as "now they are going to have to deal with an up swell of very committed and very alarmed conservationists."

"What is amazing is how willing the state is to destroy those 140 acres just because there happens to be a lot of good land there. If 100 or 140 acres of high quality prairie just like this turned up anywhere else in the state, we'd be going berserk to get the money to protect it," the DNR official told Husar. "Let us hope the skirmishes are quick and effective as Gov. Jim Edgar certainly doesn't wish to leave office with an ugly blot upon his impeccable environmental record," Husar ended.

Several years after Husar wrote the column, a movement was started to name a scenic lookout at Mississippi Palisades

State Park, two miles north of Savanna, after the noted outdoor writer. The improved lookout has not yet been built.

Arlen Dahlman, who oversaw the site selection as the Army's base transition coordinator, told another Tribune writer, Peter Kendall, that there were no Brownsfield sites at the depot available for a prison. "Plus, time wise, some of that area will take quite a while to clean up."

A short time later, the Tribune in an editorial wrote that Edgar should heed the pleas of environmentalists and intervene to move the prison to a less destructive location. "What is difficult to fathom is why the state selected a 140-acre site at the foot of Hanover Bluffs Nature Preserve with a vast prairie below. For starters, the proposed site is about five miles from the nearest developed area, and to bring roads and utilities to the new prison will cost an estimated $14 million. Building on empty land certainly is cheaper and faster than demolishing the Army base and cleaning up the contamination. But those shortsighted economies would damage part of a unique ecosystem. They also would jeopardize potential development of two job-creating industries: A maximum-security prison and an eco-tourism business attracting birdwatchers and weekend naturalists and their families who would come to enjoy one of the ever-scarcer stretches of unspoiled Midwestern prairie."

Whitney in his "Of Cattails & Cornsilk" column on June 24 complained of misstatements and lies being written about the prison siting. (8) The site was chosen after a long and arduous negotiation process between the U.S. Army, the LRA, IDNR and U.S. Fish and Wildlife Service.

"It is an agreed-upon site which minimizes the amount of damage that will happen to the significant flora and fauna which do, indeed, exist on the 13,000-acre depot."

Whitney thought the notion that the land was virgin

and untouched simply was not true, as it had been farmed for years. In addition, Whitney knew that the James' Clammyweed existed in numerous other places on the depot site and believed it would continue to exist on the prison site just as the various owls and eagles, "which do not roost in that area anyway."

Disagreeing with Whitney was Howard A. Learner of Environmental Law and Policy Center in Chicago, who wrote that the army was taking the cheap way out and depriving "us all of a vanishing corner of our state's ecological history." People are growing more unhappy about the "callous attitude of state and federal officials who have ignored requests for consideration of an alternative site to protect the prairie. So why should we care about some plants and animals anyway? The sand prairie is rarer than rain forests and is part of Illinois' natural heritage. To some it might not seem like a big deal that hundreds of birds including bald eagles and threatened species of owls call the prairie home or that the James' Clammyweed is about to be gone forever," summarized Learner.

From the information he obtained from the IDNR Division of Natural Heritage, frequent letter writer Thomas Robbe of Savanna learned that only four endangered species live within the prison site. "Three of those species, the great egret, pied-billed grebe and western hognose snake can be found throughout the depot and the fourth, the Blanding's turtle, can be moved," he wrote.

The Clinton, Iowa Herald pontificated in an editorial in May that the prison would impact the river town 20 miles south of Savanna, although not much has been said about the prison there. (9) During May the Herald ran another editorial.

"Experience both in the Midwest and at other facilities

around the country has shown that parents, wives, girl-friends, children and other acquaintances visiting maxi-mum-security inmates typically relocate (if they're going to do so) to nearby communities - not to the closest little town, but to the mid-sized city within driving distance. That would be Clinton (a scant 20 miles from the prison) or Dubuque (35 miles away) - not Galena, Hanover or Savanna. The arrival of parents, wives, girlfriends, children and friends is just the tip of the impact iceberg. The effects of a prison in Savanna will be felt not only by local social service agencies, but by Clinton's retailers, businesses and industries, schools and law enforcement officials - among others," the Herald printed.

The Savanna prison is an issue "that must be dealt with throughout the area - by Savanna, Bellevue, Maquoketa and especially Clinton. Hopefully that will happen with every involved organization - from the Clinton City Council on down - sooner rather than later ... like just before the prison throws open its doors in three years. In this instance many area communities can learn a valuable lesson from their younger members. The Boy Scouts of America live by a mot-to that every one of these organizations should adopt - and quickly when it comes to the Savanna prison. Be prepared," the Herald wrote.

Years later, Robert Knuth of Savanna, who attended the first LRA meeting and most of the meetings, wrote to Lawfer saying that LRA members were supposed to be placed on the board to help create jobs, but three Jo Daviess County mem-bers, Rachuy, Wehrle and Rutherford "were anti-prison and involved in various organizations with anti-prison agendas." (10)

"When it looked like the prison would be a reality, it re-ally showed that they'd done their home work. They passed out petitions with some of their organizations to vote down the prison. The Nature Conservancy showed their face with Mr. Drucker as their point man, objecting to the prison location at Whitton Gate. After the government had spent $435,000 for clean up of all possible contaminants, the race was on. The DNR biologist, Randy Nyboer, found the James' Clammyweed on the property and stated that it was rare endangered plant. We later learned the weed is a common plant in the West and was brought here on the hooves and in the intestines of animals," Knuth revealed.

The Clammyweed seemed to be the main ammunition to thwart the future of the prison. "At the close of one LRA meeting (I attended all but five since LRA's inception) Jim Rachuy shook his finger at John Sullivan, the LRA chairman, and said, 'You will never get the prison at the Whitton Gate. I will guarantee that.' No LRA member should be making that statement - they should be looking for jobs instead," Knuth stated.

Later, Knuth went to a meeting in Elizabeth for Jo Daviess County residents to express their opinions about the prison. "A man from Galena told me not to enter the building if I was for the prison. Later, the same man gave a speech regarding why the prison wasn't good for the area. I later took the time and drove Massbach Road toward Elizabeth and stopped and talked with 21 people native to the area. All but three said that they would consider the prison as a positive step to provide jobs and that it would not harm the area. The other three basically said it is like religion - they would prefer not to discuss it, but would be satisfied if it was brought to a vote in the county."

Knuth believed Edgar didn't have the "intestinal fortitude" to stand up to the Nature Conservancy. "Later, we learned that he would be threatened with a battery of lawyers and a legal fight that could last four to five years, eclipsing the timing of the prison due to time restrictions on the funding and bond measures .... If the prison would have succeeded at the Savanna Army Depot with all the amenities and improvements, 2,000-plus people would have benefited ... Some of the people who were anti-prison from Jo Daviess County have since moved away to other areas, as if they had accomplished their goal of stopping the prison and no longer had a vested interest in the area."

Gubernatorial candidates join the fray

Chapter 11

A major stumbling block to securing federal and state grants to fund infrastructure improvements for the prison occurred in May when officials had trouble coming up with local matching funds. On May 6, at a joint meeting of the finance committees of the Jo Daviess and Carroll county boards, neither county board chairmen, William Ritenour of Carroll County or Judy Gratton of Jo Daviess County, were willing to commit their counties to providing the funds to obtain the grants.

"I can't see my board excited about giving someone else taxing authority," Gratton was quoted as saying in a Carroll County Review report on May 13. "Since the information has been received, it is premature to attempt to make a decision at this time," Ritenour declared. Ritenour recommended a joint county board meeting scheduled for the next night be canceled until the information could be presented in an organized fashion. Until more information was available, Duane Olivier, the Jo Daviess County administrator, told the committees he could not make a recommendation.

The major concern was water and sewer treatment facilities needed during the construction which was scheduled to start in October or November in 1998. Deep wells were at the depot, but three miles of pipe were needed. In addition, the depot's sewer system was designed in 1945 with a capacity of 300,000 gallons per day. The system was running at only ten percent capacity. After inspecting the system, Paul Hartman, superintendent of the Savanna Public Works Department, recommended it be scraped, as it would be too costly to upgrade.

The Economic Development Administration of the U.S. Department of Commerce can only award grants to those with taxing authority. The LRA does not have that authority, so it needed a taxing authority body to provide the matching funds in the form of bonding or a USDA loan. The LRA hoped the two county boards would agree to provide the matching funds with the provision to charge IDOC with user fees after the construction began. The loan structure would be self-sustaining with no financial liability to the counties.

Haring earlier told a County Review reporter that the grant money was crucial, but at this point no one was panicking over the threat of losing the prison. Haring and other officials were gathering information from various consultants which could be presented to the county boards and committees.

Neil P. Stechshulte, economic development director with the Blackhawk Hills Resource, Conservation and Development Agency which served northwestern Illinois, agreed to write the grant application for the LRA.

The beginning of June 1998 brought some good news for the prison supporters. The Illinois General Assembly set aside $1.2 million for depot construction. Grants of $600,000 apiece were to go to Jo Daviess and Carroll counties.

"This shows the commitment of the state to the redevelopment of the depot," explained Lawfer. It also represents a partnership between federal, state and local governments. Lawfer and state Sen. Sieben made the grants their top budget priority. It would cost about $8.7 million to build the infrastructure for the new prison and IDOC planned to cover about $500,000 of the cost.

With the state grants to the counties, the local portion would drop to about $1.8 million. Jo Daviess County Board member John Rutherford feared the costs would run above estimates. "There usually are some sort of surprises. This could be like the Frentress Lake gambling casino in East Dubuque. It could become déjà vu all over again. Every month there seemed to be additional cost," Rutherford exclaimed.

Yet another environmental group, the Illinois Endangered Species Protection Board, entered the fray on June 9. (1) The board's chairman, Dr. Lawrence A. Jahn, wrote Edgar that he was distressed "one of the state's last remaining parcels of sand prairie, housing an assemblage of rare or endangered species, will be fragmented and degraded. Surely there is a better alternative for a prison."

Edgar indicated during a stop in the Quad Cities, that

he would rethink his position on the prison. (2) "We'll take a look at it and see what the concerns and what the alternatives might be ... If we think they are valid, we will deal with that," the governor told reporters. Edgar did not want to rush to a judgment in selecting the prison site. "I felt an agreement had been reached between those who like to see economic growth and those concerned with the environment."

Because the U.S. Justice Department had to give final approval for the prison site, a Friends of the Depot member Jean Gray, told the Dubuque Telegraph-Herald that her group sent the department a copy of a letter written to Edgar opposing the prison site. The group also sent a copy of another letter of opposition written by the Illinois Natural Preserve Commission to Edgar.

On June 29, Galena Mayor Dick Auman wrote a letter asking Rep. Lawfer to lobby for a different site for the prison. (3) "A recent article in the Dubuque Telegraph-Herald by Becky Sisco stated that you had no plans to lobby for a different site for the proposed prison at the Savanna Army Depot, as, 'nobody has asked you to do that.' I would like to encourage you to do that. Moving the prison to a Brownsfield site in the southern portion of the depot makes a lot of sense, and it would please people on both sides of the prison issue. There is an opportunity to meet the economic and environmental needs of the area with a move to a Brownsfield site."

Also in June, the land to be transferred to the U.S. Fish and Wildlife Service was given the name "Lost Mound Sand Prairie." The name originated from the geographic feature identified as Lost Mound.

Pressure on Edgar continued to grow when Republican gubernatorial candidate George Ryan sent Edgar a letter on July 1 saying the prison site poses a "significant ecological impact to this pristine sand prairie." (4) "Moreover, the Environmental Impact State, Illinois Natural History Survey, Nature Preserves Commission and Endangered Species Board strongly call into question the suitability of constructing the prison in the middle of this rare, unfragmented natural area which is a habitat for many threatened and endangered species." Ryan ended his letter with a handwritten note saying "Jim, I really need your help," and signed "George."

Ryan's opponent in the November 1998 election, Democrat Glenn Poshard, issued a press release July 3 saying that he opposed Edgar's prison site selection. Poshard proposed using previously developed property to save Illinois prairie and backwater woods. "I would respectfully request Governor Edgar simply to stop now and allow the next governor to select the site and build the prison," Poshard wrote. "I support the need for a prison on the site of the Savannah (sic) Army Depot. But instead of destroying this great natural habitat, I will build the prison on the site of the abandoned buildings - already developed land ... Instead of the structure dictating the land use, let the prairie and the natural beauty of the site dictate the development." If Edgar pushed ahead, Poshard promised to use his congressional office to demand an environmental impact statement, which would delay the project until the next governor takes office.

The Sierra Club stepped up their pressure on Edgar on July 1 when they wrote him stating that the depot site is home to the James' Clammyweed and to the grasshopper sparrow, which nests in the area and is protected under the Migratory Bird Protection Act. (5) "Species such as the fragile prickly pear, the Indian paintbrush and the flame flower are found on the depot's prairies and nowhere else in Illinois," the club wrote. The Sierra Club was pleased Edgar was planning to view the area and consider alternative sites. "We welcome your reconsideration and urge you to select an environmentally and economically sound site."

Edgar was still sticking to his guns concerning the prison site, as reported by the Chicago Tribune on July 2. (6) "At this point I have no plans (to change the proposal). But ... we'll listen. We'll take a look," Edgar told a reporter. Edgar had countered environmentalists by saying that everyone was in agreement on the use of the surplus property and that "the area designated for the prison would be used for commercial development," the Copley News Service reported. While environmentalists said there was an alternative Brownsfield site, Edgar believed it was not viable.

Copley reported Edgar had not made a decision whether to locate the prison elsewhere in the state. "We do know that they (environmentalists) asked us to look at the other site. We looked at the other site. It is not viable. It is just too expensive and it's too time consuming. We need to move ahead. That site (Area H) doesn't work," declared the governor.

Edgar's comments came on the same day the plan was roundly criticized by environmental groups speaking before a citizens committee that advises the state's Department of Natural Resources. The Department of Natural Resources Advisory Board, a citizens group appointed by the governor, met in Quincy and heard from representatives of eight environmental groups including the Audubon Society and Nature Conservancy. "I've never chained myself to a sand prairie, but I'm willing to try," Eugene Gray, representing the Natural Area Guardians and Northwest Prairie Enthusiasts, warned the advisory board.

Some LRA members also were standing pat on the prison. Member Joel McFadden told a Freeport Journal-Standard reporter in a July 7 article that the prison spot is better described as an over-grazed, poor-quality pasture. (7) "The prison site had 50 cows on it this morning."

The place picked by Edgar essentially was a Brownsfield site, Dahlman told the Journal-Standard. The military had scoured the area for unexploded ammunition and removed 5,000 pounds of metal, 2,000 pounds of ordnance-related materials, including base plates from WW I mortars and 3,000 pounds of farm-related material. As of the first of July, about $454,000 had been spent preparing the site for a prison, revealed Bob Knuth, a Savanna resident and Restoration Advisory Board member.

The controversial James' Clammyweed was elusive during a search at the prison site June 29.by advisory board members. Lots of rain during the summer could have been the explanation for the plant's absence.

As for claims that the prairie land was pristine, the area was homesteaded in 1860 and farmed until 1918 when the federal government bought it for an artillery test site. For the past 25 years, more than 5,000 acres, including Site No. 3, were reseeded to pasture grass and fertilized to provide forage to the 1,200 to 1,500 head of cattle pastured there during the summer months.

At a July 6 LRA meeting, the board voted to meet with all "necessary parties" to find a mutually agreeable site. Such a move would delay prison plans and jeopardize a $2.3 million grant from the U.S. Economic Development Administration for prison infrastructure, predicted LRA Executive Director Steve Haring. "We're on the clock with the EDA. We were given 30 days to put in an application of some way, shape or form. Fourteen days are gone." The possibility of losing the grant forced the board to cut back on engineering services and include only waste treatment plans in the grant application.

Sullivan told the Dubuque Telegraph-Herald that environmental groups had been spreading "lies" about just how pristine the sand prairie was. Sullivan said he thought some groups and individuals were looking for any excuse to prevent the prison from being built in Jo Daviess County. "From the beginning, they said we should put it in Carroll County."

Sullivan came under fire in a July 8 Telegraph-Herald article when Jean Gray of the environmental group Friends of the Depot called for his resignation from the LRA board. Gray charged that Sullivan had asked the Department of Corrections to reconsider its selection for a prison site without consulting the rest of the LRA board. (8)

The Environmental Law and Policy Center of the Midwest obtained copies of correspondence between Sullivan and the corrections department through a Freedom of Information request.

Sullivan, who admitted he made the request, told the Telegraph-Herald that Gray was making "much ado about nothing. The letter is no different from any other letter I would write encouraging industry to locate at the depot," he added.

During 1996, the U.S. Army offered four sites to the corrections department for consideration. On May 21, 1996, Corrections Deputy Director Karl Becker wrote in a letter that his department preferred an area designated as Site No. 1. It was "closest to existing utilities; it requires the shortest length of roadway to be improved; it minimizes the distances for support services; and it is close to a heliport area."

But, two days later, Sullivan wrote Becker using Sullivan Food stationery, asking him to reconsider. "From a master planning perspective, Site No. 3 would be the best-suited location for the prison. It offers a more open, flat terrain for construction and is closer to the Whitton gate, which the

DOC could utilize as another entrance/exit to the prison facility. On Site No. 1, there are areas of either threatened or endangered species identified, studied and managed by the (Illinois Department of Natural Resources) that would have to be dealt with, plus some asbestos-filled storage tanks that the Army maintains." Sullivan also revealed that the LRA would be willing to extend sewer and water lines to Site No. 3.

Savanna Mayor Eugene Flack, an LRA member, was not aware of the letter, neither was Rachuy. "I think we ought to be consulted 100 percent," Flack told the board. Were environmentalists to choose between sites No. 1 and No. 3, many would choose Site No. 1, responded Rachuy, who also was involved in a environmental group called Prairie Enthusiasts.

Sullivan wrote another letter which was published in the Prairie Advocate on Aug. 12. Sullivan complained that there had been false accusations made about the letter he sent to the Department of Corrections. (9) "People should know that the Illinois Department of Natural Resources played a large role in the selection of Site No. 3. A letter dated February 1997 from the IDNR office of Realty and Environmental Planning concurs with Site No. 3, with only two species, would have the least impact. In addition, the LRA board, during the re-use planning process, voted at least two different times to support the project."

Sullivan continued, "As LRA board members, our first priority should be to replace 450 jobs that are being lost because of the closure of the base. So, yes I will continue to encourage industry to locate at the depot, and if it means writing letters encouraging them to come look to see what we have to offer, I will do that. I hope that the other (LRA) members will do the same thing."

Edgar drops a prison bombshell

Chapter 12

After the first week of July 1998, Edgar heard enough gruff from the environmentalists.

The governor dropped a bombshell on July 9 when he decided to order the Department of Corrections to hold public hearings on an alternate prison site near Thomson, about 10 miles south of Savanna. (1)

Top Edgar aides had flown to Savanna the previous day to meet with local officials about the possible new site, described as 150 acres of privately owned farmland north of Thomson. Several of those attending the meeting told the Associated Press that everyone involved was in favor of the new site, including officials who pushed hardest for the depot site. "It is still premature, but the site does appear to have many of the qualities we need for a prison without the negatives," explained Edgar's environmental aide Allen Grosboll.

Tom Hardy, an Edgar spokesman, added that the switch would not be made until after public hearings could be held and environmental and geographical surveys made. Hardy explained that the governor was committed to bringing prison jobs to the economically depressed area, and had asked his aides several weeks earlier to approach local officials about finding an alternate site.

The Thomson site was part of a 1,000-acre tract near state Route 84 owned by Alliant Energy Co. The company proposed giving 150 acres to the state for the prison and then developing the remainder as an industrial site. The company also would get the benefit of supplying power to the new prison.

"I think our biggest concern in the broad-based big picture, is good solid job creation for the region," Haring told reporters after meeting with Grosboll and Corrections Director Odie Washington in Savanna. "And if it means the local redevelopment authority kind of stepping aside on this particular project and having it placed some place within our region of post, so be it." Flack thought the depot site was the best site, but "if this is an alternative that is welcome by both the county and the Department of Corrections, then we'll be tickled with it."

Dahlman wrote a memo on July 9 to the BRAC office stating that he expected Thomson, which had a population of 538, to be the new prison site. (2) "I expect this to go through, but the state is following normal procedures. I knew there was consideration for this site as well as one adjacent to the depot, but didn't think they were at all close to making a decision. At least the jobs will stay in the immediate area. Also, the depot reuse efforts got IDOC to this part of the state."

Some residents, including Lynn Kocal, editor of the Prairie Advocate in Lanark, weren't ready to give up the depot site. Kocal said "it's time to fight for not just the LRA and the authority the U.S. government gave it, but for our schools and communities, too. We'll only get back what we put into things. If the area doesn't attract employers and jobs, property taxes will go up, property values will decline as fewer people will want to send their kids to schools here," she wrote in July.

"Business will crumble and that will raise property taxes. Property taxes will have to increase just to keep our school doors open. Old water mains will break, the streets will fall into disrepair, the cities will have to spend your scarce tax money to condemn abandoned buildings. You can bet, your property taxes will go up some more," Kocal warned. She urged her readers to write to their federal and state representatives to ensure the development at the depot.

The Quad-City Times predicted on July 10 that the Thomson site was 98 percent certain to become home to the maximum-security prison. (3) "I appreciate that the environmentalists didn't want us to build the prison on a site they felt was environmentally important, even though there were buildings put there before by the Department of Defense," Edgar told the Times. He added that the Thomson site will "save a lot of lawsuits down the road. It also will keep jobs in the area and hopefully alleviate environmental concerns."

Thomson Village Board President Merrie Jo Enloe hoped that the prison would bring in new business. "We're looking at the saving of those jobs that were with the depot. We're excited from that standpoint," she exclaimed.

The acreage offered to the Corrections Department was part of 1,000 acres the energy company picked up in the mid-1960s with the intention of building a nuclear power plant. "With that not coming to pass, they have been marketing it as an industrial site," Haring revealed.

Environmentalists with the right contacts apparently convinced the U.S. Economic Development Administration into balking at paying $9 million of the $98 million cost of the prison, the Rockford-Register Star reported July 10. State and local officials fumed that the environmentalists got to the federal agency, which helps redevelop areas affected by military base closures. (4) "There were indications the EDA might drag its feet. Obviously, people who were opposed to this site got to the EDA," agreed Grosboll.

On July 23, the Telegraph-Herald reported that the LRA had run out of time to land a federal grant to help pay for infrastructure improvements for the depot. The U.S. Economic Development Administration had set aside - but not formally granted - $2.3 million in 1998 for water and waste-treatment capacity for the prison. The agency required a formal application from the LRA, based on the reuse plan for the depot, and the prison was part of the plan. The LRA has to revise its reuse plan and submit a new grant application for 1999, the Telegraph-Herald printed.

"We will be given the highest priority. The EDA wants to fund us, to help us develop an industrial park," Haring noted. But, an EDA official told Haring that its agency wouldn't help pay for the prison unless environmental issues were resolved.

Leaders of two environmental groups protesting the site denied contacting federal officials and praised Edgar for reconsidering. "I never contacted the EDA about this, but I wouldn't be surprised if someone had. Why shouldn't we play hardball? What's unfair or wrong about that?" asked Lynne Padovan, director of the Illinois Environmental Council.

Harry Drucker of the Friends of the Depot believed the Thomson site would be a welcome compromise. "The jobs would stay in the area, and the prairie stays safe," he told the Register-Star.

In another article, the Register-Star reported state officials as saying that the Thomson site would cut construction costs, and it would not alter contracts signed with area trade unions. "There's a gas main nearby and the electricity is ready to go," declared Beth Walton, an Alliant spokesperson.

The Sierra Club notified Lawfer that they were thrilled the environment would be protected by building a prison in Thomson. (5) "We appreciate your efforts in trying to find a solution that would provide both economic development and environmental protection to the region," wrote Michele Moss, assistant state field representative with the Sierra Club. Moss also enclosed information about the James' Clammyweed and the sand prairie ecosystem. She advised that drought-resistant plants such as the James' Clammyweed usually compose a mere one percent of sand prairie plant cover. "The depot sand prairie is a large high quality tract with thousands of dependent species that can only survive in this rare sand community."

In the Sierra Club Illinois Chapter's publication Lake & Prairie for the third quarter, the club crowed about a Sierra Club victory. "Much to our delight, the governor announced in July that a new site had been found for the prison ... Thanks to all of you and our environmental partners, the governor sought a new site. Even the Department of Corrections and the Local Redevelopment Authority have gone on record stating this alternative site will work. We commend the governor for reviewing the issue and opting to preserve the sand prairie and we commend you for all your hard work. This could not have happened without you," the Sierra Club wrote.

Local residents continued to voice disgust about the prison environmentalists' victory and scorned the James' Clammyweed as a villain. "I realize that environmental protection is important, but I am appalled to think that others would put a plant ahead of the economic well-being of northwestern Illinois," Larry Libberton of Mount Carroll complained.

"Savanna used to be the hub of several railroad systems; now the trains just cruise through the city. When the railroads left, many jobs left with them. We have lost other industries as well. When the Army depot finally closes, many good-paying jobs will be eliminated. The announcement of the prison was the one ray of hope our area had seen in recent years," Libberton wrote.

Louis Ranchino wrote to Lawfer that radicals such as Friends of the Depot "must be faced up to and repelled. If you were elected by a majority of the people, you will enforce our majority to build the prison as voted upon by our LRA representatives."

In another editorial, Lynn Kocal observed that the Friends of the Depot had a hidden agenda in bringing other environmental groups into the fray. "The impression is given that no matter what site is selected, the Friends of the Depot will have an objection, since they have offered no viable solution to the problems they have found. If you believe in democracy and don't want appellate or federal courts or bureaucracies giving you laws and morality that you don't agree with, then fight back. This has to do with our preservation. The prison site is not natural or uncorrupted. It is not pristine. The site is the most easily developed at the depot." In her column, Kocal included names and phone numbers of area congressmen and state elected officials to contact.

In correspondence to Lawfer, Don Crawford of Hanover wrote that local governmental bodies were undermined by the environmentalists. (6) "Now along come a few Chicago imports who want to save a weed. They may own property here, but they certainly do not make a living here. Local people may clean their houses and mow their grass, but that is about as much contact as they have with us. I realize the environmentalists know who to contact, what chains to pull, and they have a lot of money behind them. However, this is a case where money should not talk."

Crawford had spoken with many people in Hanover and Elizabeth and believed they felt "sold out by the state because of the environmentalists. The prison site doesn't adjoin the Hanover Bluffs, as it is about one half mile away with crop ground between them. Is it OK to make golf courses, condominiums, etc., for the enjoyment of the tourists and have nothing said about the environment?" Crawford asked.

Eleven members of the Carroll County Overall Economic Development Committee signed letters to gubernatorial candidates Ryan and Poshard asking them to retract their position on the depot so a prison could still be built there. (7) "We are extremely disappointed that both of you portray that you desire to represent the people of this state and actually responded to pressures from outside of the state initiated by a small minority. The Army depot is not a pristine prairie as it has been farmed, inhabited, utilized during federal ownership and grazed. Your statements were responses to misrepresentations and have resulted in considerable additional costs that we, as citizens of the state, are having to cover with our tax dollars as the Department of Corrections seeks an alternative site," stated the letters.

On July 23, Lawfer received correspondence from Francis Snider, president of the Carroll County Farm Bureau advising that the Farm Bureau continue to support the depot for a prison. (8) "The Carroll County Farm Bureau feels the Savanna Army Depot provides an excellent area for the prison siting and that it would be unnecessary to move to a new site within the county. This feeling was unanimous," wrote Snider.

A Tom Robbe letter in the July 20 Clinton Herald declared that the average Joe American who respected his flag is fed up with Big Brother in Jo Daviess County. "The native-born working men, working women, farmer or townsfolk alike are always welcome in Carroll County, Ill. - but please leave those neighbors at home. The Savanna Army Depot has housed many lethal weapons in the past. We have lived with them, we have made a good living from them and our local environment has not suffered - as many would have others believe."

'It looks like Thomson won the lottery'

Chapter 13

At a July 20, 1998 hearing in Thomson attended by about 700 people, most of the two dozen who spoke favored building the prison at the depot. If that did not pan out, the supporters urged the state to build it in Thomson. As long as the prison was built in northwest Illinois, everyone wins, most agreed.

"If it doesn't go to the Savanna Army Depot, let's take this chance and run with it," said Luanne Bruckner, a Thomson Village Board member. Haring felt that it didn't matter where the prison went. "It will bring economic stability to this county, this region and points beyond. Let it happen," Haring told the crowd. Several local residents, including Jim Ritter, believed Thomson needed the economic boost a prison could bring. "We need the jobs," he exclaimed.

Not all agreed that the economic benefits would outweigh the potential problems. Others complained about the lack of employment opportunities for their children and grandchildren. (1) One resident sent a letter which was read to the audience. "This community is small, peaceful and friendly. I, like many others, want it to stay that way," the letter said.

Citing the experience of another small community (Ina) with a prison, hearing moderator Enloe declared "Ina has not had one family of one inmate settle in town, but the economic growth has moved more tax dollars there. I have confidence in the community that we could pull this off."

Several in the audience expressed anger and frustration toward those who fought against the depot site. "It's pretty much accepted, except for a couple of outsiders, that the Savanna Army Depot is the best place for the prison. I don't think it's safe for the environmentalists to get in front of this microphone," charged Dennis Bowman of Savanna. Judging by the applause and cheers that followed, his comments struck many nerves, Telegraph-Herald reporter Becky Sisco reported.

"How many of you people have seen a James' Clammyweed?" asked Mona Bradshaw, another area resident. "How many would want to?" And Gene Bull, a 38-year depot worker added, "when they can find a weed to stop a prison, there's something wrong."

Bud Cherry of Hanover came to the environmentalists' defense. "The difficulty with finding the James' Clammyweed is that it is an endangered species," Cherry chirped.

Although Lawfer urged the Department of Corrections officials to keep the prison at the depot, he was pleased they were continuing to support a prison in northwest Illinois, even in an area which was now out of his district. "As you know the local redevelopment authority and a lot of local citizens have worked long and hard to replace the jobs lost by the closing of the Savanna Army Depot. They have tried to turn this hardship into opportunity, and I think they've done a good job. What is important to remember here, is that if the prison is moved, it fundamentally changes the long-term redevelopment plan created by the LRA. This seems awfully late in the game to be changing the playbook," Lawfer noted.

He believed that to change the plan would jeopardize future industrial and economic growth at the depot. "In the last three weeks, calls and letters to my office are running 20 to one in favor of keeping the prison at the depot." The site also had the support of local elected officials with both Jo Daviess and Carroll County boards in favor of the depot.

"It's like Thomson won the lottery and they didn't have to buy a ticket," remarked state Rep. Mike Boland, a Democrat from East Moline. Boland, whose 71st District covered Thomson, said that although he backed efforts to get the prison to Savanna, he would like to see it in Thomson if the depot site was not viable.

Bruce Bielema of Thomson was one of those opposed to a prison in Thomson. "I can't believe we can't find a place to build a prison out there (at the depot)," Bielema exclaimed.

Although some opponents cited concerns like an increase in local crime and the possibility of escapes, Tom Messer, who lived two miles from the proposed site, felt the concerns were overstated. "A prison would not bring undesirables, but a lack of jobs and economic vitality will," he argued.

Moving the prison to Thomson would kill almost any prospect to develop the depot, Helen Schamberger, a past Jo Daviess County Board member and president of the Jo Daviess Development Inc., expressed in a writing to Edgar on July 20, the day of the Thomson hearing. (2) "Without an anchor, like the prison, EDA cannot commit millions of taxpayers' dollars to the depot, and without an anchor there will be no infrastructure improvements to support other development. So far, a small group of Jo Daviess people have managed to use politics in the name of a weed to derail the LRA's plan for development of the depot." Schamberger added that nowhere on the base could the endangered weeds be found. "This certainly leads me and I'm sure it does you also, to wonder about the identification of these weeds in the first place," she wrote.

Besides the Jimmy Weed, the other environmental concern was the sand prairie. "All of Illinois is a sand prairie, and I don't see development being halted anywhere else in Illinois on this lame excuse. There should be large concerns about taking out 150 acres of farm ground if the Thomson site is chosen. We local people in Jo Daviess County have to fight a small exodus of people from Chicago who think they have found a Mecca and do not want any changes. These people don't want a four-lane highway, a prison (all in the name of environmental concerns), and even small hog farms. These people do not need jobs and, for the most part, don't vote Jo Daviess County. Local people can't afford to put weeds and bugs over jobs and grandchildren," Schamberger wrote.

On July 22, the EDA officials told Haring there would be no infrastructure funding that year for the depot because of uncertainly surrounding the prison plans. (3) The Freeport Journal-Standard wrote on July 25 that a small but dedicated group worked diligently behind the scenes to convince Edgar to move the prison to Thomson. Environmentalists influenced the governor by working with an influential law firm, advisory boards to the Department of Natural Resources, key members of the press and through personal associations they had with other environmental groups and the governor's staff. (4)

Eugene Gray, a member of the Friends of the Depot, attributed their success to being a dedicated group "with right on our side" who "felt something was going on that wasn't right." Gray said their beliefs led them to the Environmental Law and Policy Center Chicago, which used some of its $2.3 million annual budget and experience with blocking done deals - such as a proposed Chicago suburban toll road expansion - to persuade the governor to reconsider his decision.

Thomas Buchele, a staff attorney at the center and former litigation partner at the prestigious law firm of Jenner & Block in Chicago, handled the environmentalists' case. In an interview with the Journal-Standard, Buchele said that people in the environmental groups, including Harry Drucker of Wilmette and Hanover, founder of Friends of the Depot, and some private individuals now affiliated with environmental groups, had talked with staff at the Environmental Law and Policy Center in the spring.

The center agreed in April to take their case, and Buchele started investigating issues involved with the prison site. "We came to a number of conclusions. One was that this was a bad decision policy-wise in that ... it is an ecologically very important area, not only for the state of Illinois, but the entire Midwest," he noted.

Next, the center and its clients put together what Buchele called a "huge cooperative effort," which organized to advocate for a change by forming and mobilizing a coalition of environmentalists around the state. "We were able to bring together a lot of groups who hadn't worked together before," Buchele revealed. These included the Nature Conserverancy, Aububon Society, Sierra Club and Illinois Environmental Council. A former Fulbright scholar, Buchele had donated legal services to the Sierra Club and was a member of the board of the Illinois Environmental Council, which he described as an umbrella organization that lobbies in Springfield for environmental issues.

LRA member Jim Rachuy felt the Tribune editorial probably weighed heavily on Edgar's decision. "It certainly created a lot of publicity," he noted.

Although some of the prison supporters believed that someone got to officials high up in the federal government, Rachuy confirmed that the only federal officials who contacted him were legislative aides for Sens. Durbin and Moseley-Braun. He said that neither he nor anybody he knew had contact with Vice President Al Gore.

Edgar makes it official

Chapter 14

Edgar formally announced he selected the Thomson site for the prison July 30, 1998, after signaling three weeks before that the state would not build on the prairie at the depot.

The official announcement was delayed as the Department of Corrections officials needed time to make certain the Thomson site had no hidden liabilities, such as archeological relics or buried waste. To the contrary, officials had found that the new site would save the state money because it would require less earth moving and fewer infrastructure improvements. "In 17 years of reviewing sites for prisons, I have never seen a better one," declared Washington, the DOC director.

The Chicago Tribune wrote that the announcement almost closed the books on the environmental uproar which followed the April announcement that the state would build a 1,100-bed prison on the native prairie at the depot. (1)

Enloe was thrilled with the news. "Now we will have a positive industry in our community for those who may need job replacement with the closing of the Army depot, or higher paying jobs with benefits, and for a number of our residents who are looking for economic stability for themselves and their families." (2)

To a significant number of people in Thomson, the placing of a prison generated concerns. Enloe promised, "On behalf of the village board, I pledge our utmost attention to those concerns. At a minimum, we will attempt to maintain the quality of life to which they are accustomed; however, my intent has always been and is to improve the quality of life for our residents."

Cindy Hook of Thomson told the Clinton Herald that she didn't think the state gave Thomson enough time to consider a prison in its midst. "They gave us three weeks to think about it and they gave Savanna three years. We're not done yet." Hook revealed that a group of residents were going to drive to Springfield to protest the decision.

"I still can't imagine why they couldn't pursue placing it in Savanna. It's probably a good thing for the area ... but it probably should have been located at the depot," observed Robert Gray of Thomson. Enloe empathized with some of the residents' concerns. "We understand there are people who have concerns, and we plan to address those."

Some of the proposed needs included improving roads to the site and increasing the Thomson police force to some form of full-time service. All four officers in 1998, including the chief, worked on a part-time basis.

Jo Daviess County Chair Judy Gratton was pleased the prison would stay in northwest Illinois. "I'm glad it's at least in the area and the jobs will be retained in northwest Illinois."

According to the Telegraph-Herald on July 31, officials at the Environmental Law & Policy Center in Chicago broke into loud cheers when they heard that the prison would not be built at the depot. "A prison at Thomson seems unlikely to raise environmental concerns," said Howard Learner, the center's executive director. Learner added that he had not seen the site and was waiting results of tests by state natural resources officials.

The announcement did not please Sullivan. "I think it's a shame it wasn't placed at the Army Depot. I still think the Army Depot is the best spot. I wish (Edgar) would have stuck to his guns. An anchor tenant is needed at the depot," Sullivan grumbled. "I'm sorry the environmentalists had as much influence as they have had."

Sullivan told the Telegraph-Herald that the decision meant many of the prison jobs would move south. "It's still good for the region, but it's better for the southern part of the region - Morrison, Fulton, Thomson. This is going to affect Jo Daviess County," Sullivan added.

Although Edgar believed the new site fulfilled his goal of bringing jobs to the area, people driving down Route 84 a short time later saw lots of "No Prison" signs in the yards of Thomson residents. Many in Thomson felt betrayed by a prison coming to their town. Several residents told a Telegraph-Herald reporter that the facilities with 1,200 beds would overwhelm Thomson, with a population of about 550. (3) "All prisons are overcrowded. This one would probably double, which would make it almost five times the size of

Thomson," Kevin Green, a farmer who helped organize op-position to the prisons, complained. "A prison would bring more traffic and more safety risks. Prisoners could escape and their family members and friends could move into the community, spawning gang-type activity."

Opponents also were worried about the expense of pro-viding water and sewer-treatment facilities to the prisons. Green went on to explain that even if grant money covered the cost, the village could be stuck with the bill until the grant money came through. He believed that a prison would add to the workload of local law enforcement and the Carroll County court system.

According to Green, 852 people from Carroll and sur-rounding counties signed a petition opposing the prison. "We feel sort of blindsided," Green told the Telegraph-Herald, explaining that some people in Thomson resented that the governor spent so little time choosing the Thomson site. "It was first brought up July 8, the public hearing July 20 and he made his decision by the end of July."

Edgar spokesman Eric Robinson told the Telegraph-Herald that although the governor had been contacted by the prison opponents in Thomson, he had also received a lot of support for the site.

The Chicago Tribune editorialized that Edgar deserved praise for intervening and fashioning a "Solomon-like solu-tion" to a prison standoff.

Later, Sullivan revealed that he had contacted Alliant Energy officials earlier to see if they would donate the Thomson land for a prison in the event the depot proposal fell through. "I talked to (an Alliant officer) when the rumble (with the environmentalists) started. I asked him to go to Alliant and ask them if they would give us land down there just in case. He had a different spot than where it sits now that they were going to give us. Yeah, I had that lined up."

"The sad part of it is, the environmentalists who got it stopped are not even around Hanover, anymore. The Grays, Cherrys and Sturms left. The Sturms left with some buildings sitting here in Hanover. Another thing that hurt us, when we were getting letters of support from communities, Hanover wouldn't give us a letter. ... Edgar wanted it up there though. I do know that because I saw him a few years after he was out of office and I had a conversation with him. He said 'that's still where the prison should have been," Sullivan added.

Betty Shimp of Savanna expressed her disagreement in a letter to the Dubuque Telegraph-Herald on Aug. 14. (4) "Isn't a shame that the wealthy and powerful of Jo Daviess County consider the preservation of the James' Clammyweed more vital than the economic welfare of their less fortunate neighbors within Jo Daviess and Carroll counties. And, if they did not have 'Sir James Clammyweed' on which to base their claims, what other noxious weed would they be protecting? My opinion! They are selfish N.I.M.B.Y.s (not in my back yard) types. Many of these folk are transplants from crowded cities. They made it big so why give a hoot about jobs and economic stability for others," Shimp wrote.

"One of the vocal opponents to the prison at the Savanna Army Depot lives in Hanover and has constructed four very ugly brown metal buildings south of Hanover, right along the Great River Road, Route 84. These buildings obstruct the view of the lovely hills to the west. I wonder who asked for an environmental study on this property?" Shimp quiered, "Let's all collect seeds from the important weed and scatter them on the pristine lawns of those who love it so. Who's with me?"

Dahlman believed that calling Site No. 3 area pristine was inaccurate and thought the environmentalists overstated their case. "There were farmsteads there ... so all that land was farm land. I think they would have objected to any place especially within Jo Daviess County."

The Department of Corrections accepted Site No. 3 after it was shown all four sites. Dahlman thought that more area for new industry would have opened up had the $8 million to $9 million in infrastructure improvements been funded by the government. "Site No. 3 also was ideal for other types of development ... it would have maximized use of over 3,000 acres," he added. "There was a map the DNR put together identifying the plant and animal species. You could go any place on it and you wouldn't get 300 acres without something being there."

Dubuque Telegraph-Herald reporter Becky Sisco recalled that the prison story was one of the most divisive issues she ever covered in Jo Daviess County. (5) "Although I dislike sitting in hard folding chairs and listening to long-winded speeches, this event (March 2000 depot de-activation ceremony) felt different from most others I had covered. I could feel the passing of an era. Some people cried silently as the U.S. flag that had flown over the base was sheathed for the last time. The day before, several employees had received service awards for the work they did during the closing.

"For many, a new prison represented good jobs. It would replace all the jobs lost when the depot closed, and then some. It would create new opportunities for existing businesses, which would serve the workers and the families who would visit people incarcerated there. And it would leave room for plenty of other development at the depot, such as an upscale resort community along the Mississippi River," continued Sisco, who has since written a book about Jo Daviess County.

"For others, it represented nothing more than a quick

economic fix and a blemish on the otherwise beautiful, rural landscape. Sure, it would create jobs, but it would discourage further investment in upscale housing, clean industry and something called eco-tourism. They believe, if handled correctly, the transition of the property from military to civilian use could build on the area's tourism industry. And some were concerned about the impact a prison would have on native endangered and threatened species," she noted.

When the depot prison proposal was withdrawn because of the threatened and endangered species, many prison supporters felt bitter. Cisco wrote, "I heard several snide remarks about the James' Clammyweed, just one of the species being protected, with people saying that mere weeds were given more consideration than people. Many believed that rich people who had moved to Jo Daviess County from the Chicago area had used undue influence to get the prison project stopped."

By the time the prison became a dead issue, Sisco reported that she was frustrated by people on both sides of the issue. "Some prison supporters seemed so intent on making the project a reality that they didn't seem to want to listen at all to arguments from the other side. They seemed to forget about the other 20-odd species on the state endangered or threatened lists, and they seemed insulted by the fact that I reported on the wildlife issue for the TH (Telegraph Herald). Two of the leaders refused to talk with me, making it difficult for me to present their side.

"I sensed that, although many of the opponents were genuinely concerned about wildlife, some got on the wildlife bandwagon simply because they did not want a prison in their own backyard. Some did not want to let the 'wrong element' - family members and friends of prisoners, who might come from a different socio-economic class or be of a different color - into the county," Sisco wrote.

Sisco felt bad for any part she might have played in making the issue divisive. "Afterward, I wrote an article about healing and conflict resolution, but I don't know if anyone paid any attention to it. Now I worry about whether the base will ever be cleaned up enough to let the public use it to its potential. I wrote several articles about the process, hoping to get citizens interested in it and to push for action. But the only time the public seemed to show much interest was when the Army decided to close part of the slough to fishing and hunting," she concluded.

Eight to ten years later, Haring, who later ran as a Republican for a seat in the Illinois House of Representatives, was pleased that "in the end we were able to keep that facility here in our area but I'm very disappointed it wasn't constructed at the Army depot on that property. I'm disappointed it wasn't part of the Redevelopment Authority's economic development plan. It's unfortunate we had to wait this long to get long-term job creation out of it. The construction jobs were outstanding but … full employment is a goal which is yet to be achieved." (6)

Haring said that "moving the prison to Thomson was just a decision made by the governor and the executive office that they weren't going to fight the environmental fight at that particular time. "If I recall, we had the Sierra Club, we had the Friends of the Depot, we had the Illinois Nature Conservancy, we had agencies within the Illinois Department of Natural Resources that were encouraging him and everybody else not to construct there. It got to a point, I thought without asking him directly, they weren't going to fight that battle. We were able to work with Alliant Energy and the village of Thomson and quickly come up with that alternative site to keep it here in northwestern Illinois. I'll never forget the day we took a quick windshield tour of that site in Thomson," reminisced Haring. "The folks at the Department of Corrections fell in love with that and away we went."

The Department of Corrections came in early on and looked at the entire Army depot. Haring explained, "it was the dream of then DOC director Odie Washington to have an administrative facilities, training facilities as well as standard inmate housing. ... we gave them multiple options, four different sites, and they ultimately came in and decided site No. 3 was where they wanted to go. It made sense to them."

Haring believed the environmentalists overstated their case. "I'm not anti-environment; nor was anybody on the Redevelopment Authority board at that time or yet today. We were just trying to strive for a balance, a balance between nature, the environmental concerns, in this case the threatened and endangered species, and an economic development balance, the Department of Army's cleanup issues." He explained that they were ready and willing to work with the Illinois Department of Natural Resources at that time in coming up with some creative ways "to work with, work around and work through those threatened and endangering species concerns to have that particular site developed. Outside interests just kept putting the pressure on."

Although few of their plans for the depot materialized, the environmentalists, offered Haring, "are still there . . .we really haven't seen any sustained new development to this date at the Army depot. This is not to slap the Development Authority directors or anybody involved with development there. We haven't seen any new construction." Haring named the opportunities available at the site: the warehouse distribution, the rail, the telecommunications, and the igloos. Interest "has been on and off again, and on and off again. I know they had companies come to them at the LRA seeking their ground to develop. There were environmental concerns threatening endangered species. The message was sent a long time ago to don't mess with the endangered and threatened species," Haring declared.

Haring had no disrespect for the environmentalists and their charge, but thought there had to be a balance. He didn't think the depot property was pristine as they claimed, knowing that the ground was farmed before it was Army depot property. "Some say that land is untouched by human hands is a far cry (from the truth). Our families have got to work and have to live. We have to be cognitive of the environment as well and I think we can do that. It can't totally be one way," he reasoned.

Now that the sand dunes along the river and portion of the depot have become a National Heritage Corridor because of the discovery of Native American artifacts, Haring would like to see much of the depot used for parks with walking and biking trails together with well-planned, well-balanced development. "I think there is some ground available for development of business and industry. The environmental cleanup has held up redevelopment to some extent, but there is 'freed-up property' available."

Haring heard rumors circulating in northwest Illinois that former Vice President Al Gore, who had environmental concerns, was contacted by the prison opposition, but he hasn't any proof or memos written by Gore in his possession. "Rumors and things like that get going, but I've never seen a memo by Gore when he was vice president ... but it's always a possibility. There was a lot of people being contacted back then.

"A lot of powerful people politically and a lot of influential people" were involved in the prison controversy. The opposition "just took off and the whole concept of not only the corrections facility but economic development just took a major blow," he added.

"You don't have to go too far in Savanna and Carroll County to see this region needs good-quality jobs. Thomson is a $140 million facility that's ready to go, and it just amazes me they couldn't find the funding to get it open five years ago, ... knowing that other Department of Corrections facilities are tremendously overcrowded. We have staffing issues at East Moline, for example. They have close to 1,500 inmates and 150 to 175 correctional officers working 24-7. That's a terrible ratio. Dixon has 2,220 to 2,300 inmates and 240 correctional officers. We hear more and more as people are being made aware by correctional employees who are being more vocal, that their lives are on the line every day. More incidents are happening every day where guards are being beaten. We had the hostage situation in Dixon recently and a sexual assault (of a prison psychologist). These things are going on and on," declared Haring.

The prison controversy "isn't over and it's very frustrating," Haring continued. "I can go down to Geneseo on the campaign trail, in the Quad Cities and a mayor in the Quad Cities asked me 'what in the world is going on up there?' The frustrating thing is they are still dealing with the same issues we dealt with back in 1995-96. They are still struggling with threatened and endangered species. They are trying to get agencies to move a little bit to the middle to a level of compromise and balance.

"Let's be creative and open-minded here in being able to work together for the common good of all. We have to create quality jobs out there and preserve the environment. I think we can do that. Why not an ethanol plant or a biodiesel plant or why not a technology park? You don't have to go too far from here, Rockford or Chicago or Minneapolis,

… where they have industrial-commercial park development with ponds and wildlife, flora, fauna and walking trails. It's a nice balance. We have 13,000 acres out there and the LRA just wanted a small piece of it. Here we are 10 to 12 years later and we're still struggling," Haring emphasized.

But no hiking or biking trails or an interpretive center have been established yet at the depot. Rachuy, during a phone interview said that he felt this was because federal officials don't want the public there while hazardous materials remain to be cleaned up. Money for the clean up isn't a high priority with the federal government, he explained, because the war in Iraq is diverting federal funds. (7) "There still is a lot of planning going on. The depot needs to be cleaned up first."

Although he heard part of northwestern Illinois, including an area south of Stockton, was once considered for a national park, Rachuy doubted whether the depot and the scenic areas nearby would ever be turned into a national park. "Nationals parks are just playgrounds for us old folks," he chirped.

"Saying the James' Clammyweed is native to the area is like saying the Asian carp are native to the Mississippi River," declared Whitney. "If you fish all of them out, they would be endangered, too …. Saying and proving it are two different things. That's the tendency of groups, and I think it's what we all do. When espousing a particular cause, we marshal all of our forces and we all probably tend to exaggerate to prove our point. And, at the same time, we ignore the other side entirely," Whitney added.

A much broader issue than just the prison was involved and concerned Whitney. "With the government jobs at the depot, the area became dependent upon all the money these people made … When that was taken away from here … we didn't have anything to fall back on. It hurt our culture here tremendously in ways I think people don't realize. Many people who worked at the depot had a significant amount of education. Not all, but many ... contributed to the community. They belonged to the school board, the park district ... It was much like losing Shimer College leaving Mount Carroll. We lost those leaders and intellect," Whitney reflected.

Those who worked at the depot had served in foreign countries and contributed to the community in several ways. "Many of the depot people traveled the world," Whitney said, "and they shared what they learned with the community. Certainly the prison would have alleviated some of that, but not to the extent the depot was doing."

Instead of the environmentalists getting Edgar's ear, Whitney believed the cornerstone behind Edgar's decision was the federal government pulling back the EDA grant. "I always thought it was Adlai Stevenson (former U.S. senator who owns a farm north of Hanover) who pulled out the stops within the Democratic Party," Whitney reasoned. "The loss of the grant stopped it more than anything else. Whether Stevenson contacted Gore (former vice president Al Gore) ... it could have been done at a staff level, of course. Adlai Stevenson, and he still is, was something of a power within the Democratic Party. So then you get to whoever is in charge of this EDA thing, who says flat out, 'no.'"

Don Crawford of Hanover, who had been on the LRA board since its beginning, disagreed with the Stevenson scenario. Crawford insisted that the EDA grant was pulled after the federal government learned the prison wouldn't be built at the depot. He felt a lot of people from the Chicago area who had contacts in high places, got to Edgar's ear. "They knew whose chain to rattle and how to do it. They had a lot of money behind it ... There was the silent majority who didn't say much ... There were two or three people who were really instrumental in getting the prison killed. A couple of those people moved out." (8) "If the LRA had gotten the EDA grant, there would have been money to run sewer and water lines to the Whitton gate area, which would have opened it up to other development. That would have been the backbone for more industry to come in," Crawford predicted. " Then they would have upgraded the sewer plant."

When Thomson applied for an EDA grant to build sewer and water lines for the prison, an EDA spokesperson told local officials that if it would help the prison, the grant probably would not be approved. Thomson officials were to state on their application that the funds would be used to serve a Swedish-owned sweetener plant, several miles away.

"It was a real touchy thing. You couldn't mention prison in connection with it at all ... So what stopped it (Savanna depot) was the EDA grant. It wasn't going to happen. The endangered species, none of that was going to stop it. I think they hung their hat on environmental issues. But I think there was a huge political thing here that was probably the cornerstone of this whole story," Crawford explained.

Crawford noted that Thomson would not have gotten the prison, if it wasn't for the groundwork laid by the LRA in trying to get the prison at the depot. "I think the Thomson prison (whenever they get it open) will help northwest Illinois. But if it had gone in at the depot, it would have been up and running before the money dried up."

In an attempt to get the Department of Corrections to open the Thomson prison, several local leaders argued that the state should close some of its older maximum security prisons and transfer the inmates to Thomson. Whitney opposed this move.

"During the move to open the Thomson prison, some

said, 'gee, let's close Vandalia because that's an old, old prison.' I don't support that ... Vandalia's whole economy is tied
to the prison. Somebody said 'let's get all of them (the correctional officers) to move up here.' Right. They all have mort-
gages on their house. Throwing 700 people out of work. They
can't sell their house. There is no market. If they can pay for
their home (in Vandalia), they can't buy a new one up here. It
would be devastating to them and the community," Whitney
pointed out.

Whitney was on vacation in New York when he got a
phone call notifying him that Edgar had picked Thomson for
the prison. "It was quite an emotional day when that phone
call came ... it looks like it may open up (now) in a limited
extent anyway," said Whitney, adding that the inmates were
going to start being brought in after Labor Day.

The area continues to fight for economic development,
but in the 40 years Whitney has been a newspaper publisher
he said, "the scenario plays out a little differently. "We fight
for the beautiful ecological area we have. That includes the
sand prairie here, the river, the bluffs and even the farmland
out here. It's quite significant. We're trying to have all of this
and have a tremendous work force here ... Maybe what we
have to say is 'that is going to happen,'" Whitney pondered.

As for the LRA bringing in business at the depot, Whitney
believes "nothing was working and it still hasn't .... There
sure was a lot of carpetbaggers (who flocked to the depot to
claim the land)."

Although some feel the environmentalists folded their
tents and disappeared after their depot victory, Whitney
thinks they are still a significant force in Jo Daviess County.
Over the years, he has received news releases about their activities, but has refused to print them because of his continued ill feelings over what they did to the area.

When they selected the LRA board, they wanted a good mix, so they chose environmentalists, businessmen and county board members, said Sharon Cholewinski, LRA administrative assistant. (9) "However, they all had their own ideas on how to run the LRA and redevelop the depot. With that, came disagreement. John Sullivan and Steve, I believe, in my own heart, that they were concerned for Savanna and wanted to do something for Savanna. And I'm not saying the other people didn't want to do something good, but Steve and John were looking for the best things for Savanna, being it environmental or being it business, just to bring business and development to the Savanna Army Depot, which would then help Savanna and Hanover economically," Cholewinski explained.

Others saw the depot as 13,000 acres of pristine landscape used only for viewing. "We had the U.S. Fish and Wildlife Service which felt they should keep it all. Originally when they negotiated with the LRA, they talked about developing it into trails, special programs and viewing areas," she added. Nothing was done to invest in eco-tourism features because the USFWS and the Illinois Department of Natural Resources didn't have the money for it.

Mesquaki Tribe representative Preston Duncan worked feverishly trying to get the Indians together. "They just couldn't get their act together. They were divided amongst themselves. Preston had wonderful ideas. Whether they were feasible, was another question," Cholewinski wondered.

In his role as an LRA member, Bob Wehrle of Galena was also anti-prison. Rachuy worked for the environmentalists, the Eagle Watch group, headed by Rutherford, and several other groups. "Everything John Sullivan and Steve (Haring) brought to the board ... with all the networking and contacts they made ... the board asked 'should we do this? It seems like we are giving everything away.' Nothing is worth anything unless someone is willing to pay the price for it ... The board kept rejecting all of these proposals," Cholewinski said. "Nothing was right to them. It either wasn't enough money or we were giving too many buildings away. They nitpicked every proposal ... and there were some proposals which were not good ones. A lot of deals have been made with people who moved into the depot and went bankrupt. But that comes with the territory when you're trying to develop a new business."

A label-making company in Savanna planned to relocate to the depot to expand its business. "He wanted a particular warehouse, but he only had so much money to invest ... because he needed special equipment for his expansion. They pooh-poohed that idea," she exclaimed. "A winery was interested in locating at the depot, but Bob Wehrle didn't want it because he said that there were so many wineries already in the Galena area. LRA member Penny Lauritzen of Lanark brought in many agricultural-related proposals, but "because they were agricultural and didn't bring in a lot of money and required some buildings and the bunkers, those ideas were rejected," Cholewinski complained.

A housing development was proposed on the bluffs overlooking the Mississippi River, but Rutherford didn't like that idea because housing was available at Galena Territories, and "who would want to build near Savanna,?" responded Cholewinski. Rutherford had added that the housing would be disruptive to the eagles and that idea was "deep-sixed."

"It was frustrating because Steve worked very hard to make contacts and bring in things ... but everybody would be very negative about it. Don Crawford was a positive (board member) and Don Schaible (mayor of Hanover) had to walk a fine line, because a lot of people who protested the prison lived in Hanover," Cholewinski noted. "But it was frustrating to get Jo Daviess County to work with Carroll County. Carroll County really wanted to push on the prison and Jo Daviess didn't want it."

Later, the Jo Daviess County environmentalists on the board started "getting on John Sullivan's case and Steve was becoming more and more frustrated. You could see he wanted to say things, but felt he couldn't." Cholewinski revealed that finally Steve decided he had taken the LRA as far as he could take it, and he resigned. Some time later, Cholewinski learned in a public meeting that her job was being eliminated. She said that she could apply for a secretary job at half her salary, but resigned instead. Later, she learned the money saved from her job was earmarked as a higher salary for the new executive director, John Morehead of Galena.

A former LRA board member, Morehead indicated that he was for the prison, but Cholewinski produced an anti-prison petition showing where he had signed his name. Cholewinski had been told to make her minutes of the LRA meetings "more vague." She was not surprised to learn that one of Morehead's moves, when he took over, was to shred much of the LRA files and papers.

Sullivan said his duties as LRA chairman took two years out of his life. "We didn't even take vacations, something my wife still reminds me," he said. Taking a break from bailing hay on his farm during August of 2006, Sullivan said that many from Hanover who protested the prison have moved away.

$200 million spent and two-thirds done

Chapter 15

A factor greatly affecting reuse of the depot concerned the environment. Hazardous materials were dumped at various sites. The principal contaminants were TNT, solvents and unexploded ordnance. In 1979, the U.S. Army Toxic and Hazardous Materials Agency identified 59 potential areas of concern during an initial assessment. An environmental survey between 1982 and 1988 led to listing the depot with the U.S. Environmental Protection Agency. The U.S. EPA, state

EPA and the depot signed an agreement in September of 1989 to investigate the sites and remedy the problems. Thirty sites were addressed in a 1994 preliminary report and by 1996, 176 sites were listed in a environmental restoration program report. The first sites to be cleaned up were a washout lagoon and fire training area. Also, underground storage tanks and PCB-containing transformers were removed.

The Agency for Toxic Substances and Disease Registry released a report on Jan. 19, 1989, alleging potential health concerns because of possible human exposure to contaminants in the ground water, surface soil, sediment and air. Agency representatives visited the depot on June 5, 1991 and identified several environmental problems including food-chain contamination. (1)

Beneath the depot lies a shallow aquifer. The groundwater below the depot property flows generally west toward the Mississippi River. On the southern portion, a groundwater divide exists in the aquifer, with some groundwater flowing east to the Apple River and the rest flowing west to the Mississippi River. Groundwater flow may temporarily reverse when the Mississippi is flooded.

Roland Unangst, a retired safety officer at the depot, was called back several times to work for a contractor who cleaned up the lagoons at the bomb washout plant. (2) "It's a shame we have so much contamination at the depot. It was a proving grounds with all the firing. Some times they missed their targets," he disclosed. The TNT removed from the bombs at the washout plant was sold to a Canadian company at 37 cents a pound. It was probably the same company from which the Army had purchased the TNT (at a much higher price), because all of the TNT was imported from Canada, Unangst disclosed.

Storage tanks containing radioactive contamination were

discovered at the depot and cleaned up under the supervision of the Nuclear Regulatory Commission, the EPA said. Mustard gas also was stored at the depot until it was destroyed at a facility on the north end of the depot.

Depot commander Lt. Col. John Tarpley called attention to news reports in 1983 which named the depot's waste dump as among the worst of all the military sites. "The Savanna Army Depot is not in the most hazardous category of waste. It's one of ten where significant efforts have been made to clean up hazardous wastes." (3) Tarpley denied the wastes were related to nuclear wastes or chemical gases and agents. "They're not raw explosives laying around waiting to be stepped on. They're water contaminated with organic chemicals associated with waste operations," he insisted.

Tarpley explained that in the past - more than 11 years ago - the depot disposed of its TNT wastes in six "washout lagoons," which at the time was an accepted procedure of waste disposal. The system allowed the waste products to go into the lagoons after only preliminary filtering. Residues then seeped from the lagoons into the groundwater. The Army has found traces of the compounds in wells 300 feet from the lagoons flowing toward the Mississippi River, which is one and a half miles from the lagoons. However, Tarpley said that test wells drilled closer to the river haven't shown any contamination. "Migration is quite slow and it poses no immediate hazard to the river or to any other off-post facility," he added.

The Army is not under the jurisdiction of the EPA in handling waste, but it keeps the EPA informed as it does with similar state agencies, explained Jim Nieb. A spokesman for the Letterkenney Army Depot, Chambersburg, Pa., which supervised the depot clean up. Nieb remarked, "Is there a hazard to the public? I don't believe so."

In a related development, a U.S. House subcommittee chairman Rep. Mike Synar of Oklahoma charged in 1983 that the Defense Department's system of disposing hazardous wastes generated on military sites invited fraud and could threaten public health. Two companies hired to dispose of defense wastes were under federal indictment for improper dumping of the wastes, while at least two others were under criminal investigation, Synar told the Associated Press. (4) He added that the Pentagon seemed to care little about what happened to the wastes once they leave a military installation, usually relying on the contractor's word that they were properly disposed.

Since 1978, when 59 potential contaminated sites were found, the U.S. Army Toxic and Hazardous Materials Division worked on a program to identify and rid the depot of contaminants. (5) In spite of those efforts, the depot was added to the Superfund list of hazardous waste sites in 1984. In the past, federally owned sites have not been included on the list because they were ineligible for Superfund money. However, in 1984 the EPA decided to add the sites to foster public awareness, EPA spokesperson, Judy Beck explained. The sites were listed for the danger they pose to human health and the environment.

Mustard gas and other explosives buried at the depot have a high potential for migration in the surface water and groundwater, said a June 1996 study by Ordnance/Explosives Environmental Services of Gainesville, Fla. Research in 1982 concluded significant contamination at six washout lagoons was flowing toward the Mississippi River, but no contamination was found in the backwater. In September of 1988, environmental monitoring began of 30 wells and nine surface sites at the depot. Results determined the contamination found in 1981 had not significantly reduced.

Contracts were signed with several environmental clean-up firms and cleanup of the washout lagoons was reported to be finished in Oct. 14, 1993. Four Seasons Environmental, Inc. of Greensboro, N.C., incinerated 25,000 tons of soil in the fire training area. Weston Services Inc. incinerated more than 17,000 cubic yards of soil contaminated with explosives. Tests showed the contamination had been leaking into the groundwater and traveled only about a quarter mile in 30 years. The June 1996 study said that the amount burned could have covered a football field by a depth of nine feet!

Closing the depot meant that money for the cleanup would have to come from the Army's base-closure fund instead of the Superfund, said depot environmental coordinator John Clarke. (6)

All the contamination at the depot could scare away developers. "Business people who are thinking of doing something will think 'Oh yuck,'" feared Kerstin Krippner, executive director of the Jo Daviess County economic development agency. Local officials would have to teach potential businesses that the words "Superfund site" don't necessarily mean leaking chemical barrels. "You have this vision of what a Superfund site looks like, and when you go there you don't see it. You see the prairies and the rolling hills and the eagles flying overhead," Krippner described

During 1995, depot officials estimated 20 percent of the base could be contaminated. The sites ranged from mounds of old shell casings to soil contaminated by TNT and other explosives. One area, already cleaned up, contained so much TNT that experts feared the ground could explode under certain circumstances.

Cherry told the Dubuque Telegraph-Herald in February he feared the Army was moving too fast to demilitarize the depot and would leave dangerous unexploded ordnance there. (7) He feared the property would be turned over for civilian use before it was adequately cleaned of contaminants.

Used as an ammunition dump, the coordinator of clean-up efforts estimated the work could take 35 years to complete and cost more than $340 million. In addition, an environmental impact statement concluded unexploded ordnance might be found across 90 percent of the base. "The Army wants to cut and run as quickly and cheaply as possible. If the Army cared about the environment it wouldn't have caused almost a half billion dollars worth of damage," Cherry concluded.

Clarke stated "our first priority is to deal with anything that poses a threat to human health and the environment. Our second priority is to take into account the Local Redevelopment Authority's wishes to transfer some sites before others." Clarke believed a lack of response on part of the Army put much of the clean up on hold. "We can move only as quickly as we get authorization," he told the Telegraph-Herald.

At a hearing in March, the public had little to say about an environmental impact study compiled by the U.S. Corps of Engineers for the depot. (8)

Terry Ingram, a member of the depot's Restoration Advisory Board and president of the Eagle Nature Foundation, suggested that for every proposed reuse of depot property, an environmental impact study be done. "I feel we shouldn't rush into this. I'd like to see this take another six months. I don't care how this land is developed, but I want to see it done correctly and not see kids getting sick off of this land," Ingram said.

"The Army should explain how it intends to protect the public from polluted land that will be turned over to the U.S. Fish and Wildlife Service for recreational use," said Ed Britton, district manager of the Upper Mississippi River

Wildlife and Fish Refuge.

About $200 million had already been spent, and work continued during the summer of 2006 to clean up more. Before depot property can be transferred to the USFWS and LRA, both the federal and state environmental protection agencies and the governor have to certify that it as clean property. The Army currently doesn't have the funds to clean up all the property. Large buoys warn boaters to stay out of part of the Crooked Slough because live shells are still being found there. These shells were fired from 75 mm and 155 mm guns during testing years ago. (9)

"There was demolition and a burning ground (in some of the slough area). There still is a lot of kick out where ammunition would blow up and kick out a live one on the ground and under water. They did fire some live rounds ... they know how many were fired, but they are finding some duds even along the River Road (which is closed to the public). Demolition pits were also uncovered during sweeps, so there is quite a bit of area that needs to be cleaned up and it will take a while," Dahlman predicted.

The cleanup effort has resulted in interesting discoveries. Once a site is cleaned, earthworms and every animal from the bottom of the food chain on up, including bottom-feeding fish, are examined for contamination. At some of the sites shells were found, and workers used a gravel separator running across a conveyor belt to recycle the brass from those shells. The contractors also discovered that groundhogs had burrowed into the area and made it their home.

Between August 1995 and January 1996, a demonstration project was conducted to evaluate the feasibility and cost of sifting ordnance-related debris, including unexploded ordnance from soils. Seven acres were excavated to remove debris above the water table. Separation of the debris recovered

1,578 live ordnance items, ranging from small arms (1,164 total items) to 155mm projectiles (two items). Other items encountered included mortar rounds, projectiles, rifle grenades, rockets, hand grenades, land mines, and fuses.

The excavation recovered about 620 tons of debris and approximately 19,300 cubic yards of soil. The soils were later determined to be significantly contaminated. Contractors removed 15,500 tons of contaminated solid waste and 450 tons of hazardous waste from the bottomland area, which were disposed in a permitted offsite landfill. Other activities included an ecological risk assessment, which is currently under regulatory review. Additionally, an investigation to more adequately assess the threats from ordnance and potential chemical weapons disposal is ongoing.

A pesticide burial site was identified in the northwestern part of the facility. During the 1950s, approximately 800 tons of di-nitro-ortho-cresol (DNOC) were buried in a trench and covered with soil northwest of the ammunition storage area. The pesticide burial trench was found, and the EPA determined that the groundwater had been impacted by the contamination.

A 1998 Clinton Herald report stated that more than 100 tons of pesticides were buried at the depot. During the time the pesticides were dumped, Terrance Ingram, president of the Eagle Nature Foundation, observed that the bald eagle population "crashed." During two months of the year, the depot has a higher concentration of eagles than any other area on the Mississippi River. "Whatever pollution gets into the water will affect the eagles," charged Ingram, who is also co-chairman of the Depot Clean Restoration Advisory Board. He had urged the Army to rescind its plan to close the depot because it would be cheaper to keep it open rather than clean it up. If the ground was not cleaned up, it would remain in the hands of the Army. (9)

A Dec. 12, 1999, public health assessment by the U.S. Department of Human Services pointed out that contamination runoff is unlikely but can't be ruled out. "Due to season reversal of shallow groundwater flow and unpredictable migration in fractured bedrock, monitoring of wells is urged to ensure the safety of drinking water," the report warned. The publication also acknowledged that pesticides were found nearby in the Apple River, but the source was undetermined.

The Army was licensed by the Nuclear Regulatory Commission to store ammunition containing depleted radium. Building A-1602 at the depot stored 30mm cartridges containing depleted radium, a U.S. Army report dated Jan. 27, 1997 revealed. After the ammunition was removed, a radiological survey performed during the week of Nov. 18, 1996, found no evidence of radiological contamination. Not surprisingly, the Army determined in the report, that the NRC license would not be included when the land is transferred.

Storage tanks containing radioactive contamination were cleaned up under the supervision of the Nuclear Regulatory Commission and the Illinois Department of Nuclear Safety. According to the EPA, the cleanup was completed in September 2000 and resulted in the excavation and offsite disposal of approximately 26,000 cubic feet of thorium-contaminated soils.

The government spent more than $200 million cleaning up the contamination since 1978 and another "$60 million to $80 million is needed to finish the job," said John E. Clarke, Army environmental coordinator. The cleanup for site No. 3, where the prison was proposed, cost $1.4 million.

The depot could contain as many as 176 contaminated sites and 7,000 acres of unexploded ordnance, said Ed Britton who is the Savanna District Director for the U.S. Fish and Wildlife Service Upper Mississippi River National Wildlife and Fish Refuge. Britton informed the Carroll County Board on Feb. 22, 1999, that at one of the sites, since 1952, the U.S. Department of Agriculture had dumped 840 tons of pesticides. "It is leaching down right into the ground water," he declared.

Britton also was concerned about the 120 acres of heavy metals, that were buried in the depot flood plain. He is worried the Army may restrict the land rather than clean up the pollutants. If the Army refuses to clean up the depot, Britton believes that the Fish and Wildlife Service can "manage critters," but will not be able to attract birding enthusiasts and other eco-tourists to the depot because of denied access to the contaminated areas.

All closing bases compete for funds, and Britton feared the Savanna depot will not get enough money to deal with the pollution. "We've had fat years and we've had lean years. About $10 million being spent this year. We've been told 2007 will be a lean year. That's because of the way the budget process works …. There's no indication the Iraq war is taking money away from the cleanup. It is impacting us when it comes to getting people's attention to make decisions, because they are thinking more about the war than cleaning up the base. When we need to get to generals to approve things, it's hard to get to them some times," Clarke exclaimed. (10)

Generals in the Pentagon don't come to the depot. They send their representatives out instead. In the depot headquarters, three Army employees work for the Army administering the cleanup. Two or three Corps of Engineers employees help with the contracts. In August of 2006, four different contractors with about 30 people worked on the cleanup. Three of the contractors specialized in looking for unexploded ordnance. The fourth contractor sampled soils and groundwater.

"We're coming to the end (of investigating 200 sites) and we have some cleanup to do as a result of those investigations ... the cleanups will include removing contamination from the soil, cleaning groundwater ... we have to clean up some old landfills, some old dumps," Clarke added.

Opening up River Road to the boat ramp at Crooked Slough won't happen soon because recent results of a survey along River Road haven't been good. "We've found several live items out there, which indicates we have more work to do before we have the confidence to open it up to the public," Clarke said. "Contractors last summer found three mortar rounds from 1918 time frame along River Road. The rounds were configured to hold mustard gas, but they didn't hold the gas used during World War I. They were loaded with sand or plaster."

Another site to be opened is the Primm's Pond area. Owned by the IDNR, it should be ready by August 2007.

When the depot was actively used as an Army installation, employees were free to enjoy the outdoors. Certain areas were open to hunting for employees and their families.

"But when it comes time to say this land is ready to transfer to a new owner, we look at it with a different point of view." Clarke explained that when transferring property, they have to make sure that it is not unsafe. "A good example is, before we were announced for base closure, our cleanup efforts were on environmental sites with the potential for the contamination to migrate off post. Since we were announced for base closure, we had to change our focus to say we were transferring a piece of property that was clean or as clean as it's going to get," Clarke said.

Army contractors conducted a radiation survey during the early years of the cleanup. "We had a specialized contractor in here doing instrument testing of all the storage locations of our old landfills. Then, we had a site on the east side of the installation where monazite sand was stored. The monazite sand contained ore of thorium. The government had a big tank of it stored here, but it was shipped out before I got here in '87. When they shipped it out, they had some spillage on the ground and this radiation survey identified the spillage on the ground. We did a small cleanup to get rid of it. All of the special weapons storage units were identified to have no radiation contamination at all," Clarke added. Eight to ten buildings have been identified to have contamination and may have to be torn down and removed to clean up the contamination. "We are about to award the contracts to decontaminate those buildings, and one method to remove the contamination may be to remove those buildings," explained Clarke. "These include the old washout plant. The lagoons around the washout plant were cleaned up in 1992 and 1993."

In 2003, lands and waters that were once the depot became the Lost Mound Unit of the Upper Mississippi River National Wildlife and Fish Refuge. Most of Lost Mound is fenced off to the public because of contamination as a result of manufacturing, testing and storage of munitions and explosives in the area. "Precautions are necessary to protect the public while visiting the refuge due to environmental contamination and remnant explosives. Environmental clean up is ongoing, thus some areas are restricted to the public," a visitor's guide warns!

Clarke is responsible for cleaning up the mess, but he chooses not to point fingers at those who made it. "Most of the contamination was created by people who were trying to win a war." The son of a World War II veteran, Clarke believes the pollution is a by-product of a "noble cause."

Thomson prison stays empty for five years

Chapter 16

Workmen finished the 1,800-bed, $140 million prison in Thomson in 2001, but the prison sat empty for five years. The state ran out of money to open it!

Edgar was succeeded by Republican George Ryan who could not find money in the budget to open the prison. Rod Blagojevich, a Democrat, promised in 2002 that he would open the prison if he were elected. Blagojevich was elected, however, he backed out on his promise. The white elephant continued to stand empty, much to the consternation of local officials who were hoping the prison would mean jobs for 524 correctional officers.

Lawfer was happy the prison got built in northwestern Illinois, but was dismayed that it hadn't opened yet. "If a site at the depot was selected, there would have been money in the budget to operate it, and it would have been open the past six years," he declared in an interview. (1) Although he wasn't involved in the day-to-day negotiations to get a prison in Savanna, Lawfer had met with Edgar in an attempt to convince him to select the depot location for a prison.

"Edgar realized a prison would replace jobs lost by the depot closing. Basically the decision rested on the unemployment. That was the one issue that could fill 400 jobs. The prison could equalize that out. The site looked real well. They (the environmentalists) said it was a pristine area and I wrote a letter (to the governor's staff) saying it wasn't. They came back and asked if there was another area. That's when they (Alliant Energy) got in touch with John Sullivan (to offer the land near Thomson for free). He (Sullivan) knew about that land and that it was available. That was after the decision was made that the prison would not be at the depot. Thomson came up in the spur of the moment," Lawfer explained.

In addition, Lawfer felt that the environmentalists had to get somebody at the highest level in the federal government to back down on the $8 million to $9 million EDA grant for the prison infrastructure improvements. If the prison construction hadn't been delayed by objections of the environmentalists, Dahlman believed that it would have been operating the past five years. "That's what is unfortunate," he said. "It would have been up and running and the economics of the area would have improved, much better than what it is today."

While workmen were finishing work on the Thomson prison in 2001, the Savanna Army Depot was busy closing its doors. In September of that year. Major Joseph Tirone was the 38th and last commander. Tirone was sorry people were losing their jobs. "There's nothing fun about that," noted Tirone, who along with his wife Paige and two sons were the only people living on the depot at that time.

In the spring of 2006, after the state legislature budgeted $1.2 million to cover the costs, the Department of Corrections made plans to open the prison An additional $6.7 million was earmarked for preparations and operations in the budget which began July 1. "When the prison fully opens is unknown," Corrections Director Roger E. Walker told reporters in May of 2006. "The prison's future depends on the legislature, which annually appropriates funds for the department. We could make it a medium or maximum. Anything is possible; it depends on the agency's needs." (2)

On July 24, 2006, IDOC named Frank Shaw as the new prison warden. He had been warden at Hill Correctional Center in Galesburg. Shaw was the third warden appointed for the prison scheduled to open in September with 75 employees. Cadets, who would be among the new guards at the prison, began training June 26. Other officers were to be transferred. The 15-building prison, with 750 employees, occupies nearly 150 acres near the Mississippi River.

A café waitress reported she met the new warden and heard that 14 transferred guards are now at the prison getting it ready to open. For a period of time, the prison operated with only one employee. A van of inmates came up once a week from the East Moline Correctional Center to help the lone engineer flush the toilets and do some minor maintenance.

The loss of the depot jobs and nationwide economic

downturn pushed the county's February 2003 unemploy-ment rate to 10.7 percent. It was no wonder residents again felt violated.

"People have a good work ethic around here. They just need jobs to go to," Freddie Preston of Savanna, who had hoped for a job at the prison, told a reporter.

Enloe resigned as mayor of Thomson in April. The next month, the Department of Corrections decided to open the Thomson prison. Enloe said she would have resigned earlier if that's what it took to get the prison opened. (3)

She remembered that she thought it was a great day when Edgar announced Thomson as the prison site. The village had earlier applied for the juvenile prison. As for the maxi-mum-security prison, Enloe said when they applied, "we had no idea what you had to do. Then they came back and said 'do you want it,' which was great."

Race was the main reason for the objections to the prison believed Enloe. Many of the people against it were Chicago transplants living along the river and in the bluffs. She said that they had moved away from crime to a generally white area. "They didn't want it because they were afraid of the families who would move into the area to follow the prison. They were concerned because of racial issues. I never heard this from the people who lived here the better part of their lives. I never heard them express an issue about race." Enloe knew of at least three families that moved away because of the prison, and of a farm that took a whole year to sell.

As for environmental concerns, Enloe believed there was only one objection from a Native American, who was con-cerned that sacred burial mounds would be disturbed. An environmental assessment found no mounds in the area and the man withdrew his objection.

Enloe doesn't expect Thomson to see a huge increase in

population when the predicted 700 employees are hired to run the maximum-security prison. "If we got four to eight families in here, that would be typical. I look at it from a regional standpoint. It's not just Thomson who suffered when the Army depot closed, it was Savanna, Mount Carroll and the whole area. So having the prison will impact the whole area. No one community is going to see an influx of employees. It's going to spread out all over." Enloe also noted that it could impact West Carroll School District, now that Savanna, Mount Carroll and Thomson are in the same district, because each would get more state aid with additional students.

Improvements to Thomson were funded by $11.5 million in grants and loans and used to build a new sewer plant, water plant and 210,000-gallon water tower to serve the prison. An access road was upgraded to an 80,000-pound road and an intersection was built on Route 84. The ICC also funded an improved railroad crossing. The village was never late making bond payments, as the state reimburses the village. Every year, the village has to ask for the funding to make the bond payments. "The water and the well alone will take care of the prison. It had to be done that way. If anything ever went wrong, the prison will have use of the water and the well and the village would be shut off from it. But the village has its own," Enloe explained.

The village improvements did allow more development into Thomson. However, the town has stagnated during the past six years with the prison closed. Several businesspeople made improvements to their businesses, gambling that an opened prison would bring more people in. Wayne James had to sell a BP convenience store and gas station at a loss and the old Watermelon Café had several changes in ownership because the prison remained closed.

As for the depot, no big employers have made their home there, and few jobs have been created over the years. The EDA never awarded the depot a grant to build infrastructure for an industrial park. The LRA, whose members can be found bickering with each other at many of their meetings, is still in business trying to attract employers. One LRA chairman told the Carroll County Board that his main activity was refereeing the fights between the board members and the few LRA tenants.

"Part of the problem is that we are in the backwaters. We are so rural. We are too small," explained Randy Nyboer of the Illinois Department of Natural Resources.

Depot redevelopment makes slow progress

Chapter 17

The LRA received an EDA grant to make access road and sewage improvements on the Carroll County site of the depot. Development at the depot is going slow because title of the land cannot be conveyed until it is cleaned up. Without ownership of the property, potential developers are unable to borrow money using the land as security. So far, only Rich Stickle of Cedar Rapids, Iowa, bought property for warehousing and one other depot tenant operates a railroad switching business. The LRA is, however, considering proposals for a security training business and a data farm

"There have been a number of different entities where there has been a lot of talk and a lot of promise, but they didn't come to fruition. A new company taking over for the data farm ... they are moving along quite well. They are going to use the igloos in the special weapons areas. There's not enough igloos there for them so U.S. Rep. Don Manzullo got the U.S. Fish and Wildlife Service to lease igloos in their area for 25 years. The idea for a backup makes sense. They want to put it in a less vulnerable area than a building in Chicago. They do backup storage in different places, many times in industrial parks," explained Dahlman.

The government is making it tough for CyberVault to open Bunkerfarm, a data storage business as it does not permit fiber optics to be run under the Mississippi or over the Bellevue dam to the depot, according to Unangst. A direct path along the Great River Road, Burlington Northern railroad and the Mississippi also is not available. During an Aug. 2, 2006 LRA meeting, the board voted to pay $153,282 to extend the hair-thin strands of glass from Savanna to the depot, which was renamed Eagles Landing. Cyber-Vault agreed to pay the rest of the $459,864 cost. (1)

"Bunkerfarm" consists of a series of thick-walled, earth-sheltered buildings used to securely protect paper and electronics files containing government and private information. The bunkers previously housed munitions. Illinois Information Management, a company allied with CyberVault, owns some of the structures, and it plans to lease others from the LRA through an agreement with the U.S. Fish and Wildlife Service. Many of the bunkers are at Lost Mound, a fish and wildlife refuge.

As of August of 2006, 88 people were working for businesses at the depot, most of them at Rescar Co., which does maintenance on rail cars. The company subleases from depot tenant Riverport Railroad.

The LRA has applied for a foreign trade zone designation to benefit companies operating within the zone at the depot. A foreign trade zone allows companies which are importing goods from outside the company, to defray or reduce duty tariffs which are imposed on foreign goods. If the application is successful, it will be the third such zone in the region, along with Rockford and Rock Island. All three zones plan to work together to market their facilities to companies shipping products from Asia for Midwest distribution, according to Jack Koster, president and CEO of Riverport Railroad.

Unangst complained that the U.S. Fish and Wildlife Service claimed all the river front and had fenced it off from people who could have bought depot housing. "I know several people who said they would have bought the homes built in the 1960s if they could have access to the river. They (the U.S. Fish and Wildlife Service) won't knock the fences down because they want to protect the wildlife. They have signs every 300 feet, saying 'Keep Out.'"

Disappointed that the Department of Corrections didn't build a prison at the depot, Unangst claimed that one of the objectors didn't want a prison at the depot, because its lights would interfere with his view from the bluffs where a shell fired from the depot had landed on his property. The objector didn't say much, according to Unangst, but the next week a "For Sale" sign was placed on his property. When it sold, the man and his wife moved back to the Chicago area. Unangst saw many people at the Hanover post office airmailing letters to the governor objecting to the prison. "Just about all of them were Chicago-area transplants," he noted.

Don Crawford who retired from the depot in 1991 at age

55 and farms near the depot, is discouraged by everything that happened since he joined the LRA board in 1995. "I kind of would like to see something go some time. There's been an awful lot of hurdles to jump through - government hurdles, EPA hurdles, environmental hurdles."

Three environmentalists on the board, John Rutherford, Bob Wehrle and Jim Rachuy, shot down all the proposals for development at the depot because "they wanted it to be a big park," Crawford declared. "They were against everything because they didn't want to harm the plants and animals." The absence of a four-lane highway has also held back progress.

"We have the Burlington Northern Railroad, one of the best railroads in the United States, goes right by there, but they don't want to stop for any thing ... They are not receptive to shipping stuff in and out of there ... they like their unit trains," explained Crawford. The Burlington does stop for the Rescar rail cars.

While most states would be in the middle of things helping an area like the depot develop, the state of Illinois has been sitting on the sidelines and has not helped the LRA securing grants, Crawford stated. "We needed a grant to help route those fiber optics ... and the state is backing out. Jim Sacia (a Republican state representative from Pecatonica who succeeded Lawfer along with Sieben) is on our side, but the other powers to be are just leaving us up here by ourselves and forget about us."

Asked to sum up the prison story, Rutherford, an Apple River farmer, said that "reason prevailed." He added that realistic people felt the prison should have been built on the depot's lower post where there already was infrastructure because it would have cost too much to run sewer, water, gas and electric lines to a prison located five miles away.

The turning point in the prison story certainly wasn't

the James' Clammyweed. Rutherford thought it was people realizing there was no infrastructure in place for the prison and the cost would be too much to construct it. He blamed "bad planning and politics for the prison being shuttered in Thomson for six years," and agreed with others who said Thomson was a much better site. Rutherford is not bitter with those who opposed him on the board. "I have no regrets," he asserted.

The current LRA chairman, former judge John Rapp, believes efforts to create jobs at the depot have been frustrating because "for three steps forward, we have to take four steps back." One of the biggest obstacles was interference by the Illinois Department of Resources. "They want the depot to be one big nature preserve ... while it's our mission to replace jobs lost when the depot closed." (2)

Other detriments to depot development include its isolation, lack of a fluid economy and distance to a four-lane freeway. "Its location made it an excellent place to store and clean ammunition," but not to develop a business," Rapp observed. Also, the IDNR placed restrictions on the land so the natural habitat could not be disturbed. Sacia introduced a bill exempting the LRA land from some of the IDNR interference, and the bill passed and is pending in the House, Rapp pointed out.

The LRA had planned to give land to Sauk and Fox tribes, but their proposals failed. Six to eight groups had come in and talked about starting ethanol plants, but none had made any headway, Rapp reported. "Everybody comes in with ideas, but we have the responsibility to see they have the financial capability to do what they have planned ... all we want is jobs. It's been very frustrating and it's been a difficult job (being the LRA chairman)."

Rapp disclosed that IDNR doesn't trust anybody else to

clean up the land but the Army, who spent more than $10 million in the depot cleanup last year.

Sullivan complained the government has spent "an awful lot of tax dollars" to create only 88 jobs at the former depot. "The prison should have been up at the Army depot ... We knew, if we got the prison, we would have had natural gas there. They would have run it in there for nothing. We also would have had the sewer and water and power lines. The big deal is there is no natural gas up there. So that's a big downfall now. If the prison would have been there, we would have had that now," Sullivan added.

Former Galena Mayor Dick Auman, who later ran for Congress, believed Thomson to be a much better spot for a prison. During the prison fight in Jo Daviess County, Auman had announced that the tourist industry, law enforcers and the courts opposed it along with the environmentalists. He said, "the state needs to come up with other ideas and stop locking up so many people for minor infractions, such as possession of marijuana."

Ingram admitted he was involved in the prison fight, not so much to save the prairie, but to head off the potential of an environmental disaster. He figured that extending sewer lines five miles and building a sewer lift station could create the possibility of a nasty pollution spill. "If something would have happened, there would have been hell to pay. I was not so concerned with the endangered species, as the others."

Wildlife, and perhaps endangered species can be seen in the Lost Mound area along the Mississippi River. The opening of the road allows motorists to drive to a lookout and enjoy a scenic view of the Mississippi River along with a seven and a half mile dune system rising 70 feet above the river. This natural system is Illinois' longest on the river. At least 175 bird species have been observed in the uplands at Lost

Mound. A number of characteristic sand prairie reptiles are found at Lost Mound, including the ornate box turtle and western hognose snake. Resident mammals include deer, rabbits, badgers and western harvest mouse.

Motorists driving the wildlife road to the overlook, can see wild turkeys and deer and many dilapidated Army buildings which need to be demolished. Using materials donated by the federal government, the overlook was built by Douglas Buchan of Arlington Heights, who was working for an Eagle badge, and by his troop. One of the Boy Scouts' parents was an architect and drew up the plans. A plaque at the site commemorates their effort.

Visitors to Lost Mound can now ride bikes from the lower post, three and a half miles to the overlook on the River Road. Further access is limited because of the Army's request of USFWS to "keep it that way," said Alan G. Anderson, USFWS refuge operations specialist. (3)

Another public area to be opened is the former Coast Guard boat ramp. Contamination research is being completed for the Army in that area. Once that report is received by the Army, it will be submitted to the U.S. and state environmental protection agencies, who must sign off on it before the ramp is opened. Renovated in the 1970s, the ramp remains in good shape. Anderson reports that he is working to expand the parking lot there. The ramp could be used by outdoor enthusiasts wanting to fish in the Crooked Slough, part of which was opened to the public several years ago.

During late August, a gate had blocked access to a road leading to the ramp. When the road is opened, the gate will be moved another three miles past the ramp. "The road is in good shape, but I don't think it will be open even next spring. My first priority is to get that boat ramp accessible to the public. I would like to open it up to bank fishing, too," hoped

Anderson, who believes that the area includes the best beach of all the USFWS land.

Anderson is working with the state to have a 16 and under youth deer hunt on former depot property in the fall of 2007. Trained individuals will accompany the young hunters to help them set up their blinds.

Title to about 3,000 of the 9,854 acres acquired by USFWS has been acquired from the Army, and ownership of another 1,200 acres is expected to be transferred this year. Although Anderson agrees that the cleanup is going slowly, he has noticed that a lot was cleaned up the past summer. "At least, the Army is moving forward with the research," he remarked.

The USFWS was busy cleaning up its property too. Four buildings, three railroad loading platforms and a railroad loading dock were razed and removed. In addition, 500 tons of scrap was recycled. The Riverfront Railroad has removed 12 miles of track. "Eventually all the buildings will be demolished, but that may take 100 years," explained Anderson.

Even though some of the prison supporters think the environmentalists have abandoned the depot after their victory, Anderson knows that many members of the groups have helped with the cleanup and remain interested in the activity at Lost Mound. "We have had volunteers from the Natural Area Guardians, the Audubon Society, Prairie Enthusiasts and Friends of the Depot. We hear from them all the time," he confirmed.

Randy Nyboer of the Illinois Department of Natural Resources also reports that environmentalists are active at the former depot, volunteering for bird counts and seed and plant collection. "It's kind of hard to make plans when you have to depend on what's being done (with the clean up) … we should be thankful the depot is still on the Army's radar scope and things are still being done," Nyboer said.

In December of 2005, birders from across the state flocked to the depot to participate in the Audobon Society's Christmas bird count. (4) Birders from Chicago to Champaign helped cover a 15-mile circle along the Mississippi River, counting the endangered and threatened species which are known to take refuge on the riverfront area. By noon, six novice and several serious birders logged red tail hawks, chickadees and Lapland Longspurs, as well as snow buntings, which follow the snow. After the day was over, the groups documented 41 species, several birds short of the 50 to 60 seen during most years.

The Christmas bird count started in 1900 when Frank Chapman suggested that instead of shooting birds, they could on Christmas day create another tradition - counting birds. The results of the 106th count were turned over to the Audubon Society, which tracks the longest-running database in ornithology.

The Natural Area Guardians have asked to sponsor a five-mile run and a three-mile walk on a Saturday during 2007. The route for the run and walk would start at the parade ground, go to the observation deck, then back to the headquarters building with the walk ending at the overlook. The Dubuque and Rockford run clubs plan to help manage the fund-raising event.

Once the contamination is cleaned up, Anderson believes there are great opportunities (for public access) at Lost Mound. Another barrier to public access at Lost Mound is the Burlington Northern-Santa Fee tracks. The Burlington will allow vehicles to cross its tracks only at gated crossings.

Besides managing the refuge, the USFWS has founded studies, including a four-year one to study the fragile prickly pair cactus. Another study involves researching the seed bank in soils and how the plants have responded after the grazing was stopped. The infamous James' Clammyweed is part of this study. The results may determine whether grazing should be started again to control the vegetation since the USFWS forbids burning. "Grazing isn't out of the question. It's a management tool," Anderson said.

An Illinois EPA newsletter announced that public access to most of the depot will be restricted for a long time. The Army decided to close most of the former depot to the public because a study in 2000 showed the contaminated areas pose "unacceptable risks to the public." While definite plans for future recreational uses have not yet been made, the USFWS and IDNR hope to have facilities that will allow the public to bird watch at Primm's Pond and to provide an interpretive area near a hiking trailhead.

Today, recreational fishing takes place along depot property in the Crooked Slough, Apple and Mississippi rivers. Before the depot was opened to the public, fishing in the Crooked Slough was restricted to depot employees. A variety of wildlife species is found on the depot and hunting is allowed by permit.

The depot is home to 47 endangered and threatened species and eight eagle nests. Last year, more than 700 of these majestic birds were admired in a one-day count. Eagles favor the depot because it lacks human activity. As nature takes over more of the depot, more species will have a chance to survive.

Notes

Chapter 1

(1) Booklet Savanna Army Depot Activity, published by the U.S. Department of the Army. (2) "1917-2000 Savanna Depot Activity End of an Era." (3) Army Depot publication. (4) Chicago Daily News, April 15, 1943. (5) Dubuque Telegraph-Herald, Aug. 6, 1944. (6) Dubuque Telegraph-Herald, Nov. 25, 1954. (7) Savanna Army Depot newsletter Dep-O-Gram, Oct. 22, 1971. (8) Interview with Carrroll County Review Publisher Jon Whitney. (9) State Rep. Ron Lawfer memoirs. (10) U.S. Army Material Command Ammunition Center publication, June 1974. (11) Carroll County Review, Dec. 8, 1982.

Chapter 2

(1) U.S. Army Defense Ammunition Center and School Feb. 28, 1995, fax to state Rep. Ron Lawfer. (2) Gov. Jim Edgar Feb. 27, 1995, news release. (3) Savanna Times-Journal, March 7, 1995. (4) Sterling Daily Gazette in a March 1, 1995. (5) U.S. Sen. Carol Moseley-Braun letter April 13, 1995. (6) Quad-City Times, April 14, 1995. (7) Dubuque Telegraph-Herald, June 28, 1995. (8) Quad-City Times, June 24, 1995. (9) Quad-City Times, June 27, 1995. (10) Quad-City Times, July 6, 1995. (11) Carroll County Review, June 28, 1995.

Chapter 3

(1) Freeport Journal-Standard, May, 1996. (2) Dubuque Telegraph-Herald, July 15, 1995. (3) Quad-City Times, July 20, 1995. (4) Don Crawford March 25, 2005, letter to Lawfer. (5) Quad-City Times, Aug. 25, 1995. (6) Interview with Steve Haring, former LRA executive director currently running for the Illinois House of Representatives, Aug. 6, 2006.

Chapter 4

(1) LRA Board Chairman John Sullivan copy of his testimony. (2) Illinois Department of Corrections Oct. 10, 1995 news release. (3) Copy of state Sen. Todd Sieben Oct. 12 letter. (4) Prairie Advocate, Oct. 18, 1995. (5) Quad-City Times, Oct. 15, 1995. (6) Illinois Times, February 1996 issue.

Chapter 5

(1) Rockford Register-Star, Oct. 24, 1995. (2) Savanna Times-Journal, Oct. 24, 1995. (3) Savanna Times-Journal, Oct. 24, 1995. (4) Freeport Journal-Standard, Dec. 21, 1995. (5) Galena Gazette, Feb. 14, 1996. (6) Jim Rachuy letter as president of Northwest Illinois Prairie Enthusiasts, Jan. 27, 1996. (7) Illinois Department of Natural Resources press release, Feb. 27, 1996. (8) The Nature Conservancy of Illinois letter, March 6, 1996. (9) Carroll County Review, Feb. 28, 1996. (9) Judy Cherry letter in Carroll County Review, March 13, 1996. (10) Interview with former LRA board chairman John Sullivan, Sept. 1, 2006. (11) Chicago Tribune, May 28, 1996. (12) Expected Impact of Savanna Correctional Facility, July 15, 1996. (13) Galena Gazette, July 24, 1996.

Chapter 6

(1) Carroll County Review, June 12, 1996. (2) Northwest Illinois Prairie Enthusiasts letter, Jan. 27, 1996. (3) Savanna Times-Journal, Aug. 20, 1996. (4) Carroll County Review, Aug. 21, 1996. (5) LRA Executive Director Steve Haring to Brent Manning, IDNR director, Aug. 21, 1996. (6) Carroll County Review, Aug. 28, 1996. (7) Prairie Advocate, Sept. 4, 1996. (8) Savanna Times-Journal, Sept. 10, 1996.

Chapter 7

(1) Galena Gazette, Nov. 13, 1996. (2) Freeport Journal-Standard, March 7, 1997. (3) Savanna Times-Journal, July 24, 1997. (4) Prairie Advocate, Aug. 20, 1997. (5) Savanna Times-Journal, Nov. 6, 1997. (6) Maximizing the Economic Benefits of the Expansion of the Upper Mississippi River Wildlife and Fish Refuge of the Savanna Army Depot, prepared by Clarion Associates of Chicago for the Friends of the Depot and the Nature Conservancy of Illinois.

Chapter 8

(1) Quad-City Times, Dec. 7, 1997. (2) Tom Robbe letter to Gov. Jim Edgar, Jan. 27, 1998. (3) Freeport Journal-Standard, March 12, 1998.

Chapter 9

(1) Carroll County Review, April 8, 1998. (2) Dubuque Telegraph-Herald, April 9, 1998. (3) Dubuque Telegraph-Herald, April 15, 1998. (4) Nancy Winter letter, April 15, 1998. (5) Freeport Journal-Standard, April 15, 1998. (6) Dubuque Telegraph-Herald, April 19, 1998.

Chapter 10

(1) Thomas Robbe letter in Clinton Herald, April 20, 1998. (2) Dubuque Telegraph-Herald, April 9, 1998. (3) Illinois Nature Preserves Commission minutes, Aug. 4, 1998. (4) Dubuque Telegraph -Herald, June 8, 1998. (5) Illinois Natural History Survey scientist Geoffrey A. Levin memo, May 26, 1998. (6) Illinois Endangered Species Protection Board June 9 letter to Gov. Edgar. (7) Chicago Tribune, June 9, 1998. (8) Carroll County Review, June 24, 1998. (9) Clinton Herald, May 15, 1998. (10) Robert Knuth letter to Ron Lawfer, April 2, 2005.

Chapter 11

(1) Dr. Lawrence A. Jahn, Illinois Endangered Species Protection Board letter, June 9, 1998. (2) Dubuque Telegraph-Herald, June 9, 1998. (3) Galena Mayor Dick Auman letter to Rep. Lawfer. (4) Rockford Register-Star, July 5, 1998. (5) Sierra Club letter to Edgar, July 1, 1998. (6) Chicago Tribune, July 2, 1998. (7) Freeport Journal-Standard, July 7, 1998. (8) Dubuque Telegraph-Herald, July 8, 1998. (9) Prairie Advocate, Aug. 12, 1998.

Chapter 12

(1) Quad-City Times, July 9, 1998. (2) Army base transition coordinator Arlen Dahlman memo to BRAC office, July 9, 1998. (3) Quad-City Times, July 10, 1998. (4) Rockford-Register Star, July 10, 1998. (5) Sierra Club letter to Rep. Lawfer, July 9, 1998. (6) Don Crawford letter to Rep. Lawfer, July 17, 1998. (7) Carroll County Overall Economic Development Committee letters to gubernatorial candidates George Ryan and Glenn Poshard. (8) Carroll County Farm Bureau letter to Rep. Lawfer, July 23, 1998.

Chapter 13

(1) Dubuque Telegraph-Herald, July 21, 1998. (2) Helen Schamberger letter to Gov. Edgar, July 20, 1998. (3) U.S. EDA Economic Development Administration letter to LRA Executive Director Steve Haring, July 22, 1998. (4) Freeport Journal-Standard, July 25, 1998.

Chapter 14

(1) Chicago Tribune, July 31, 1998. (2) Clinton Herald, July 31, 1998. (3) Dubuque Telegraph-Herald, Aug. 29, 1998. (4) Dubuque Telegraph-Herald, Aug. 14, 1998. (5) Dubuque Telegraph-Herald reporter Becky Sisco letter to Rep. Lawfer, May 25, 2005. (6) Interview with former LRA executive director Steve Haring, Aug. 7, 2006. (7) Phone interview with LRA member and Northwest Illinois Prairie Enthusiasts president Jim Rachuy, Aug. 9, 2006. (8) Interview with LRA member Don Crawford, Aug. 10, 2006. (9) Interview with LRA administrative assistant Sharon Cholewinski, Aug. 10, 2006.

Chapter 15

(1) Agency for Toxic Substances and Disease Registry report, Jan. 19, 1989. (2) Interview with Roland Unangst, retired depot safety officer and LRA member, Aug. 9, 2006. (3) Rockford Register Star, Aug. 17, 1983. (4) Associated Press, Aug. 10, 1983. (5) Rockford Register Star, Oct. 5, 1984. (6) Clinton Herald, July 7, 1995. (7) Dubuque Telegraph-Herald, Feb. 23, 1997. (8) Dubuque Telegraph-Herald, March 7, 1997. (9) Interview with Terrance Ingram, president of the Eagle Nature Foundation and co-chairman of the depot clean restoration advisory board, Aug. 21, 2006. (10) Interview with

Arlen Dahlman, former Savanna Army Depot base transition coordinator, Aug. 3, 2006. (11) Interview with John E. Clark, Army environmental coordinator, Aug. 31, 2006.

Chapter 16

(1) Interview with state Rep. Ron Lawfer, Aug. 2, 2006. (2) Prairie Advocate, July 26, 2006. (3) Interview with former Thomson mayor Merrie Jo Enloe, Aug. 4, 2006.

Chapter 17

(1) Prairie Advocate, Aug. 9, 2006. (2) Interview with current LRA chairman Judge John Rapp, Aug. 21, 2006. (3) Interview with U.S. Fish and Wildlife Service refuge operation specialist Alan G. Anderson, Aug. 21, 2006. (4) Freeport Journal-Standard, Dec. 18, 2005.

Epilog

By Jo Anne Gale

Paul and I have driven by the prison outside Thomson often as we make frequent Savanna visits to enjoy our trailer at Seven Eagles Campsite. We camp near the Spring Lake Wildlife Area of the Upper Mississippi River Wildlife and Fish Refuge.

We have walked and biked the scenic byways of the area, observing woodpeckers, deer, otter, badger, and once a muskrat climbing a tree. The pleasant sound of honking geese, traveling in V form over the refuge, has greeted us in the early mornings. Both of us have witnessed eagles soaring in the sky above the mighty Mississippi River and have seen many unique and colorful plant species. One time Paul saw both an eagle and a turkey vulture sitting on a sandbar in the Brickyard Slough near the former Army depot. Another time we talked to a photographer working for National Geographic magazine. She was taking photos of birds along the refuge.

My camera lens captured a beautiful crane fishing in the backwaters one afternoon. I missed the shot, though, of a mother turkey and her chicks crossing one of the depot roads. Both of us chuckled as the last little wayward guy finally made it to the other side under the watchful eye of his mother. During boat trips on the river, the beauty of the bluffs has been impressive, as we steered through fields of lily pads. On visits to the lookout at the depot, Paul and I admired the view and appreciated the uniqueness of the sand dunes. We continually marvel at the beauty of the area.

We have eaten at restaurants in the area which frequently change ownership or close. When driving through Savanna, we see the closed doors of businesses and disrepair of buildings and homes. Many of the businesses still operating have for sale signs, as do many homes. The declining number of students forced three school districts to merge into the West Carroll School District. A general malaise of the downtown is obvious. We are saddened by the steady decline of a once busy and thriving area.

With Paul guiding, I have visited the depot. The presence of so much contamination is distressing, but the poignant words of Army Environmental Coordinator John E. Clarke "they are the product of a noble cause" gives me solace. Also along with Paul, I have toured the huge (empty) Thomson prison. The presence of so much barbed wire on such a tall fence, intrigued me and I aimed my camera. The guard reprimanded me, saying pictures are not allowed for "security purposes."

I wonder what would have been, had the prison not stood empty for five years!

Recently we enjoyed an afternoon at the Wildlife Prairie Park near Peoria, Ill. We saw many animals and plants all native to Illinois and in their natural habitats. Such a park could be a great tenant for the depot.

We stayed at our trailer during the times Paul met with Ron Lawfer, Tom Kocal and Sharon Cholewinski for their meetings at the Prairie Advocate office in Savanna. Although not involved in the meetings, I attended the final three-hour one and got to meet the trio of advisers in person.

The kink in my neck reminds me of the many hours spent editing this book. Paul and I worked well together - he has a gift for writing easily and fluently and mine is attention to detail. We are first timers in the book writing and publishing process. Since we are such a good team, this may not be the last.

Appendix A

The following interview was imagined by I. Ronald Lawfer with the honorable James' Clammyweed at the weed's home which Lawfer visited in August of 1999.

Who are you?

I am Polanisia jamesii or commonly called James' Clammyweed, a rare plant species in Illinois, and on the list of state endangered plant species. Unable to compete with other vegetation in the lush productive soils that covers most of Illinois, I can survive only on a sand prairie without any other vegetation and am very abundant in many of the dry, sandy western states.

An annual dicotyledonous about 12-15 inches tall, I flower in August and produce seeds before my death each fall. If the weather and moisture conditions are right, my seeds sprout in the spring, and we have another generation of James' Clammyweed.

Some people say that drought-resistant plants like me (how in the world did I ever get that name?) usually compose a mere one percent of sand prairie plant cover. Evidently I cannot stand crowded areas and am not a good neighbor.

My first public appearance in Springfield occurred when the Endangered Species Protection Board minutes of May 15, 1998, referenced a 1997 report of the Illinois Natural History Survey study saying that my home was a Class C sand prairie. The minutes go on to say that Harry Drucker of Wilmette testified that I live in an area called Site 3. This area has been called pristine, but I have lived through years of farming, artillery shell testing and even when the area was dug up in the search for unexploded ordinance.

Are there many of you around here?

"I may be a small weed, but am one of thousands that grow throughout the 13,000 acre Depot. With a specialized blowout sand habitat, I can live on the Savanna Army Depot forever. Boy, will I miss those large earth movers at the depot that regularly dug up the earth to cover the igloos! This is what gave me those needed blowouts on open areas where they found me.

Have you had any visitors?

In 1998, four people from Savanna came to visit me. The four were escorted to a spot near the proposed prison site 3, and there I was, an endangered weed living in an open sandy area. Terry Dunk from Savanna, was among the group and remembered the visit in a letter dated July 27, 1998, addressed to you (Ronald Lawfer). In his letter, Dunk asked that you visit the area to expose the foolishness of the environmentalists' claim of a "pristine sand wilderness." Dunk believed that where I live is "anything but pristine." Protecting me would cost millions of dollars in economic growth to the residents of Illinois and could cause the economic collapse of the northern Illinois communities of Savanna and Hanover.

Later, after determining that I was simply "an inconsequential little piece of vegetation," Dunk questioned my "roots." He believed that I was not native to this area and speculated that I was brought here by seeds in cattle dung from ranchers' herds that were permitted to graze on the depot. It seems the environmentalists got their way because of manure.

Dunk has not forgotten what has happened since his visit to me eight years ago. I heard that he "remains as bitter today as he was at the time the prison construction was halted." And, Dunk was convinced that the groups who opposed development at the depot did so because they did not need any economic activity for their lifestyles, and believed that growth here would make Carroll and Jo Daviess Counties more like the city they had fled. I know that Dunk is a Savanna business owner, and am not sure he will be a friendly visitor the next time he sees me.

But why would the question of you being in manure be raised?

Would it be because I am found in at least twenty counties in Kansas, two in South Dakota, and one along the Mississippi River in Wisconsin? What better way to travel than on an animal, in its digestive system, or on a cattle car of the Canadian Northern railroad, which has bordered the eastern side of the depot since the late 1800s. Cattle and sheep grazed annually on the depot to keep the grass down and reduce the risk of fire in the ammunition storage area.

My friend, Dr. John E. Edinger, member of the Endangered Species Protection Board, doesn't believe in the possibility of my seeds migrating and told me that I was "not palatable to cattle, that my seeds have no mechanism to stick to animal fur, and that dispersal by cattle appears unlikely." My spread all over the depot remains a mystery.

Since the prison is not coming here, it appears that I could live forever, if the grass does not grow too tall. I really liked it when the grazing cattle and the farm implements pushed my seed down in the sand close to any moisture so that there could be another generation of James' Clammyweed.

Are you any relation to the Northern Spotted Owl?

As recently as December 2005, I again made the newspapers. This time it was thanks to Randy Nyboer who works for the Illinois Department of Natural Resources. He was quoted as saying "James' Clammyweed is the spotted owl of the depot." Boy was he right! The Northern Spotted Owl issue was not to save the owl, but to stop the logging. The James' Clammyweed issue was not to save me as a plant, but to stop the prison construction and the 700 jobs it would have provided for northwest Illinois.

Are you a super weed?

Remember, I'm not very pretty and am short, less than 12 inches tall, but I am MIGHTY. I stood in the way of a $140 million prison.

Appendix 2

Mission accomplished, but whose?
By Thomas Kocal

Since 1995, when the announcement was made that the Savanna Army Depot would cease operations after nearly 80 years of creating jobs in northwestern Illinois and eastern Iowa, hundreds of citizens began the daunting process of creating new opportunities at the 13,000-acre site.

From the outside, the opportunity to gain access to this much land for economic development purposes was tremendous. The former federal property would be placed back on the property tax rolls, generating much-needed income for local taxing bodies.

The goal was to attract businesses that would create good-paying jobs. People would stay in the area and new families would move to northwestern Illinois. Payrolls created by these jobs would circulate into the local economy with property and sales tax revenue increasing. It was a good mission, pursued by good people, with honorable intentions.

When Gov. Jim Edgar announced that Site 3 at the depot would be the location of a new, state-of-the-art prison, these grassroots economic developers understood the ramifications of a state project of that magnitude. The prison would be Local Redevelopment Authority's anchor tenant with its infrastructure - water, sewer, gas, electricity and fiber optics improved and expanded to operate the prison. Much of the contaminated land left by the U.S. Army would be cleaned up. Because of the state's involvement, this project would create business opportunities at a much faster pace than earlier projections. Other business opportunities would include eco-tourism.

This is where the problems started. Others had a mission, too.

For example, the mission of the Nature Conservancy (TNC), as stated on their web site, is "to preserve the plants, animals and natural communities that represent the diversity of life on earth by protecting the lands and water they need to survive." Land acquisition is one of the ways by which TNC achieves its mission.

Another tool to achieve goals is a conservation easement, a voluntary, legally binding agreement which limits certain types of uses or prevents development from taking place on a piece of property now and in the future, while protecting the property's ecological or open-space valued. Included in TNC's seven core values, also found on their web site, is a commitment which states they "respect the needs of local communities by developing ways to conserve biological diversity while at the same time enabling humans to live productively and sustainably on the landscape."

Others, including the Friends of the Depot and the Sierra Club, as well as the Nature Conservancy, are not-for-profit, private organizations. They monitor the land and its uses, preserving and protecting its inhabitants, people, plants, insects and animals alike. TNC claims to respect the environment, and at the same time enable people to live off the land in a safe, productive manner.

Another example comes from the Illinois Nature Preserves Commission. The purpose of the INPC, as presented in its May 22, 1998 letter to Edgar, quotes the Illinois

Natural Preservation Act (525Il CS 30). This provides that the INPC is to "keep watch over the protection, management and the use of nature preserves…"

The act also provides the commission to "promote by advice and other assistance the protection of natural areas in the state which are not dedicated as nature preserves." Further, Section 17 reads "all public agencies shall recognize that the protection of nature preserves, buffer areas and registered areas is the public policy of the state and shall avoid the planning of any action that would adversely affect them."

To protect nature preserves, lawmakers have allowed appointed and unelected commissions to dictate policies to every public and private agency, business or individual in the state of Illinois. There is no discussion as the commissions are the "experts." and have the knowledge and evidence to make the final decisions. It is either their way or no way.

To most people, it was clearly evident that the Savanna Army Depot (SAD), a blown up, contaminated, proving ground since 1917, was not a nature preserve. The land prior to 1917, and during its tenure as a military base, was tilled, farmed and grazed on by cattle.

Yes, under the surface at the SAD, there is sand. But it was not, and is not, a "pristine sand prairie," as it was inaccurately called many times during the campaign of former Secretary of State George Ryan, when he was a candidate for governor. He called the site "pristine" in his letter to Edgar on July 1, 1998, that urged him to move the prison somewhere else. Contaminated, blown up, grazed land can't possibly be defined as "pristine." Or can it?

Yes, certain plants and wildlife are few and far between on the property known as Site No. 3, where the prison was to be built, but that doesn't mean they're endangered. That site had been cleaned up and wasn't "pristine." Was the

James' Clammyweed endangered? It was endangered when it was being shot at and contaminated by the U.S. Army. If the plants and wildlife survived that abuse, they should have survived a little construction and cleanup. Maybe the site wasn't conducive to their survival.

Site H was recommended as alternative to Site No. 3 by the INPC and was mentioned in Harry Drucker's testimony at the 159th meeting of the INPC on May 5, 1998.

Why hasn't the depot area been opened to the public earlier. Because it is a polluted brownfield, not a "pristine prairie." Drucker didn't mention some of the buried treasure on the depot is usually referred to as UXO or unexploded ordnance, possibly live bombs. Now, eight years later, the land is still considered by the Army to be unsafe for public use.

Now called Eagles Landing, jobs here are harder to find than James' Clammyweed.

Everywhere the work of the environmentalists presents a challenge to all government agencies. If the demands of these "clubs" are not met, a lawsuit is threatened, as in the case of the Prison on the Prairie. Just what is the goal of an industry, when a large contribution is made to an environmental group?

Efforts are continually made by business and industry to keep our environment clean. Environmental laws and mandates make up most of the new laws legislated in the past 30 years.

Many businesses have brought jobs to small towns and large cities. These jobs improved this nation's lifestyle and protected our borders and our sovereignty. Legal immigrants, like my grandparents who migrated from Poland at the onset of World War II, were welcomed with open arms, to work for these businesses, raising their families, and creating a lifestyle unlike anywhere else in the world. Now these same laws may be responsible for businesses moving to foreign lands, where strict environmental control does not exist, leaving behind a talented, dedicated workforce clamoring for good jobs.

It is apparent that these not-for-profit environmental groups have more power than the legislators we elect to run our government. Their actions have forced economic hardship on northwestern Illinois. Their threatened lawsuit to prevent the construction of a state prison on a brownfield - not a pristine sand prairie - stopped the economic development plans of the LRA and tied up its limited resources. And with their connections in the Illinois Department of Natural Resources and the U.S. Fish and Wildlife Service, they are continuing today.

What happened to the environmentalists' original mission? These groups were concerned about our environment and helped bring attention to air and water pollution created by irresponsible corporations. Now it's just a power play: environmentalists vs. big business. Have the leaders become so ecologically minded that they've placed their cause above the lives and welfare of the people?

The double-talk from state agencies continues to be troubling. For example, appearing in the May 3, 2006 issue of Carroll County's Prairie Advocate News, is an article entitled "Igloo users confident of success." The new company, interested in leasing and developing at least 31 J-area bunkers from Riverport Railroad, is Illinois Information Management. The bunkers will be used for "cyber vaulting," a way to protect confidential information, similar to the now-defunct Savanna Depot Technologies Corp. (SDTC).

SDTC initially proposed 2000 high-tech, high-wage jobs when the concept was first proposed in 2003. Two igloos (former ammunition bunkers) had been outfitted as marketing tools for the data storage concept. When the development was brought to LRA three years ago, the USFWS lacked the legal authority to manage a lease of the igloos on its property because a memorandum of agreement with the Army had not been signed. The Army legally leased hundreds of igloos to the LRA for 25 years. The LRA then leased them to SDTC.

Now the Wildlife Service wants more say over its property. "Our concern is the associated activities," said Ed Britton, USFWS district manager. If Illinois Information Management used the igloos, they would more than likely want the entire area closed for security reasons. "That would block any public recreational use," Britton said.

Here we go again.

Thomas Kocal
Publisher, Carroll County's Prairie Advocate News
Chairman, Carroll County Prison Focus Group
Member, board of directors, Carroll County Economic Development Corp.